Sustainable Engineering Forum 2013

Core Programming Topic at the 2013 AIChE Spring Meeting & 9th Global Congress on Process Safety

San Antonio, Texas, USA
28 April – 2 May 2013

ISBN: 978-1-62748-466-4

Printed from e-media with permission by:

Curran Associates, Inc.
57 Morehouse Lane
Red Hook, NY 12571

Some format issues inherent in the e-media version may also appear in this print version.

Printed by Curran Associates, Inc. (2013)

For permission requests, please contact AIChE
at the address below.

AIChE
3 Park Avenue
New York, NY 10016-5991

Phone: (203) 702-7660
Fax: (203) 775-5177

www.aiche.org

Additional copies of this publication are available from:

Curran Associates, Inc.
57 Morehouse Lane
Red Hook, NY 12571 USA
Phone: 845-758-0400
Fax: 845-758-2634
Email: curran@proceedings.com
Web: www.proceedings.com

TABLE OF CONTENTS

(30a) Progress and Challenges in Emerging Clean Energy Technology: A Case Study in Hydrogen and Fuel Cells from the U.S. Department of Energy

Sunita Satyapal, Office of Hydrogen, Fuel Cells and Infrastructure Technologies, U.S. Department of Energy, Washington, DC

A record $260 billion has been invested globally in clean energy, and trillions of dollars will be invested in the coming decades. According to a recent Pew report, global investment in the clean energy sector is 600 percent higher than it was in 2004. As a result of this rapid growth, the clean energy sector recorded its trillionth dollar invested in 2011. To seize this market and job creation opportunity, the United States invests in programs that advance research, development, manufacturing, and deployment of clean energy technologies. The U.S. Department of Energy's Office of Energy Efficiency and Renewable Energy (EERE) works to move clean energy technologies forward across different stages of development, from initial feasibility studies and applied research through enabling broad commercialization. A key component of the clean energy portfolio, the hydrogen and fuel cell industry has seen a more than 35% growth rate in just one year. Recent technical accomplishments include reducing the projected high volume cost of fuel cells by 35% since 2008 and more than doubling durability in just a few years. American Recovery and Reinvestment Act investments led to the deployment of more than 1,000 fuel cells for early market applications. These successful projects led companies to invest in more than 3,500 fuel cell forklifts and 1,300 fuel cell-powered emergency backup systems without any DOE funding. Interagency and international coordination of safety, codes and standards and systems analysis is critical, as well as technology validation and market transformation activities. Although the availability of wide-spread, low cost natural gas has contributed to market parity, continuing chemical engineering and chemistry challenges include the need for increased fuel cell durability, decreased fuel cell costs, and a bridge to hydrogen infrastructure.

Overcoming Regulatory Challenges In Chemical Management - Inherent Opportunities for the Chemical Industry

Jacqueline Sibblies, PE
Environmental Compliance Consulting
P.O. Box 5013, Somerset, NJ 08873
sibbliesj@gmail.com
(908) 531-4029

The first decade of this century has undoubtedly ushered in multiple changes and new beginnings in the way we live and do business locally, nationally and internationally. Several newly implemented and updated regulations and policies have resulted in profound changes in the operations of various industrial sectors. Among the industries significantly affected by new regulations and policies is the chemical manufacturing industry, uniquely subject to multiple regulatory and policy changes which, while not implemented with much public fanfare, are nevertheless sparking creative, new industry paradigms.

The industry is required to comply with extensive environmental and health/safety (EHS) requirements at local, national and global levels. Many of these regulations are administered by different regulatory bodies, and many have overlapping requirements. For example, in the United States (US), chemical companies are subject to on-site chemical handling/safety rules by two agencies - the Environmental Protection Agency (EPA) through several regulations including the Clean Air Act requiring the Risk Management Plan (RMP), and the Occupational Safety and Health Administration (OSHA) through requirements like Process Safety Management (PSM). In addition to these federal (or national) environmental, health and safety (EHS) requirements, the industry is also subject to global policy requirements such as the Strategic Approach to International Chemicals Management (SAICM) administered by the United Nations Environmental Programme (UNEP). All of these requirements include components on safe chemical handling to preserve worker safety and prevent or mitigate releases to the environment. In essence, chemical companies are obligated to comply concurrently with numerous chemical management standards by multiple entities. This is considered burdensome by most and a detriment to the business survival by many industry stakeholders.

Yet despite this perceived threat to business profitability, fulfilling these regulatory and policy requirements offers inherent opportunities for business growth and development. Indeed, some industry leaders like Dow Chemical Company (DOW) and BASF have already seized the opportunity to develop sustainability tools and models while devising a comprehensive system for fulfilling the varied regulatory requirements. In addition to addressing social, ecological and business profitability concerns, sustainability explorations could lead to discovery of alternative raw materials and applications, new product development and improvement in process operations, all positively affecting business performance. Results could extend into the global arena through the forging of strategic alliances to overcome national barriers that may inhibit ease of international compliance. Furthermore, these alliances could lead to improved global presence and established international branding. This paper will summarize key regulatory challenges in Environmental Health and Safety (EHS)/chemical management faced by the chemical industry and explore opportunities for business advancement through overcoming these challenges, citing industry cases.

Myriad Regulatory and Policy Requirements

For the past four decades, the chemical industry has been subjected to growing regulatory obligations. These have been promulgated primarily in the form of new regulations and significant amendment to existing regulations. The industry, for its part, has sought to lend its voice to the rule making process in an effort to mitigate any adverse effects to its operations that final decisions might cause. These efforts, however, have been tempered from time-to-time by various mishaps and catastrophic events which precipitate mounting public concerns, sometimes outrage.

From the 1976 Seveso disaster in Italy (which was catastrophic to livestock and other animal life), to the 1984 Bhopal India accidental toxic release (which caused thousands of deaths), to the 2005 Texas City, Texas refinery explosion and fire (which claimed fifteen lives), chemical accidents have served to frame a perception that the chemical industry is inherently dangerous. These accidents not only brought the responsible facility into high visibility and under subsequent intense scrutiny, but led to repeated critique of the entire industry. They resulted in the enactment of regulations and policies mandating preventive safety measures not only from local or national jurisdiction, but also from regional environmental regulatory authorities and international stakeholder consortia.

The 1976 incidents prompted the European Commission (EC) to issue the Seveso Directive, legislation governing prevention and mitigation of chemical accidents. The Bhopal incident generated debate in the United States Congress, resulting in the Emergency Planning and Community Right-to-Know (EPCRA) legislation, focused on emergency response and public education. There were further environmental, health and safety evaluations and discussions which led to the 1990 amendment of the Clean Air Act which required companies to develop a risk management plan. Over subsequent years, other regulatory and policy requirements effected for the purpose of having the industry practice better risk management in chemical handling. The EC and the US governmental agencies were not the only bodies which proceeded with these types of regulatory action; the United Nations (UN) also developed internationals standards. Table 1 (Appendix A) gives a summary of current regulations that contain chemical management requirements. Figure 1 illustrates the network correlation.

Fig.1. – Interconnection of National, Regional and Global Chemical Management Rules

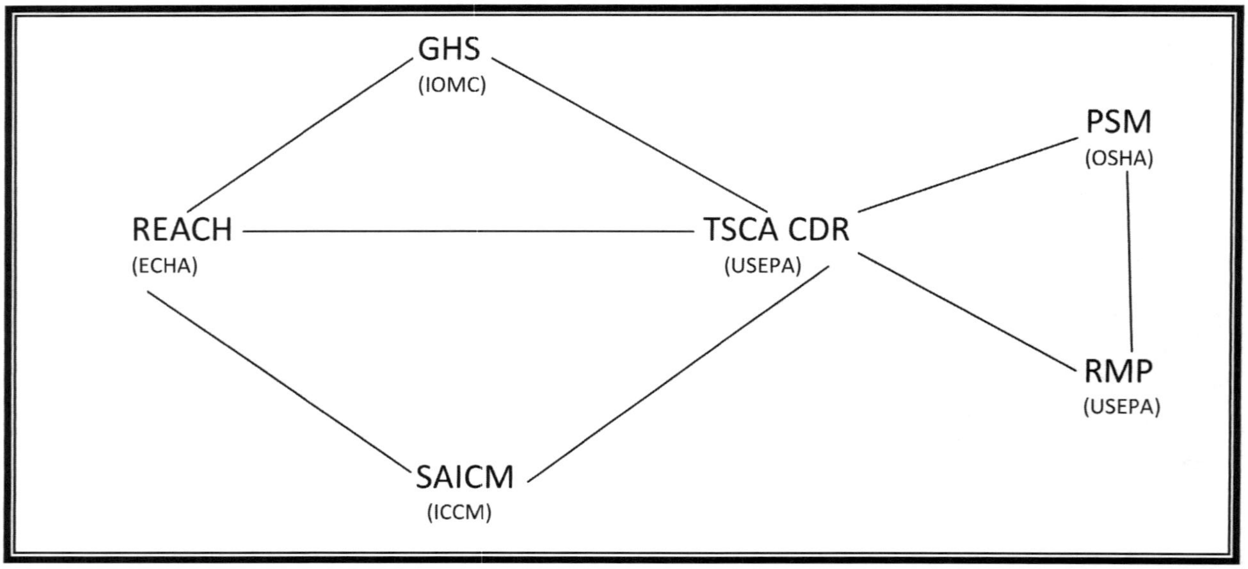

Challenges to Compliance

The Chemical industry had growing concerns about the feasibility of complying with all applicable regulations within reasonable economic limits. The financial cost of compliance was thought to be overwhelming.

For large and some midsized companies with potential capabilities for meeting compliance with in-house resources, it was felt that the cost of equipment and technological tools necessary to meet compliance would be disproportionately high. They also cited the detraction from production that additional labor-hours would require to fulfill requirements like testing, monitoring, record keeping and reporting – adversely affecting business profitability. Smaller companies, perceiving very little or no in-house compliance resource capabilities would have to rely on consulting assistance which meant more cost to a usually already lean budget.

The complexities of some of the regulations, it is safe to say, were even more disconcerting to many stakeholders in the industry value chain. As figure 1 above shows, the same agency has enacted chemical management requirements in separate regulations. Further, in the United States multiple agencies (e.g. the EPA and OSHA) have administered regulations with similar requirement for the industry. On top of that, there might also be unique state, county and city regulations that some companies have had to comply with. Small and mid-sized companies have needed assistance to decipher pertinent regulatory requirements from a multi-layered regulatory system.

To compound matters, each set of regulations established a reporting schedule, which was seldom in sync with any other set of regulations. So there was the potential for having several compliance schedules to comply with in any given calendar year. Missed deadlines could lead to a non-compliance citation, which came with a financial penalty. Companies needed a management system to manage chemical management compliance.

In addition to the direct costs associated with fulfilling regulatory requirements, the chemical industry had greater concern about the far reaching effects that these cost had on the business of doing business. A major threat to business performance in one geographic location is always competition from another geographic location. US chemical companies voiced strong opposition to many EPA regulatory requirements on the grounds that the addition cost of compliance measures added to the cost of production and that would force a product price increases, thus undermining the ability to adequately compete with companies from other geographic locations not subject to similar requirements. Alternatively, if product prices were kept at competitive levels, a company's profit margin would be undermined, posing a threat to its economic standing or even its very survival.

Compliance Strategies – Inherent Opportunities

However, demand for products from the chemical industry has been steadily rising, even through the great recession of 2008. New technological innovations brought new uses for chemicals and improved efficiency in products, creating new business opportunities for the industry. The public may ask for stringent regulations that protect its health and safety and also preserve the environment, but it also wants the highest quality and most technologically advanced products that the industry can deliver. So the question that key industry figures started grappling with was, "How can we meet stakeholder demands while simultaneously meeting the company's business goals and more importantly, shareholder expectations?"

Concurrently, as effective dates for compliance became imminent, many companies, particularly industry leaders, started taking a critical look at compliance requirements, and formulating strategies by which to achieve them. The exercise uncovered a key fact - although regulations are many and varied, requirements were found to not be mutually exclusive. As Table 1 shows, there is appreciable overlap of chemical management (and EHS) requirements not only with intra-agency rules (like the EPA's), but inter-agency rules and inter-regional and international standards as well. This scenario presented the opportunity for a company to devise a comprehensive compliance program. The chemical industry's perception of a burdensome regulatory system slowly started to change. A paradigm shift started developing around the turn of this century.

The new thinking gave rise to the development and application of environmental management systems (EMS) tools and models developed by EMS teams. Most large companies comprised their EMS teams from internal company labor pools, with team members taken from departments and divisions directly affecting chemical management and with personnel having the knowledge and experience with affected processes.

It also ignited wider interest in industry standards developed by industry and stakeholder consortium, for example, the chemical industry's own Responsible Care Initiative and the International Organization for Standardization (ISO) guidelines for environmental management (ISO 14001). With these available mechanisms, companies did not have to reinvent the wheel when exploring strategies for chemical management compliance; they already had these established platforms on which to build.

Many large, international companies had already been subscribing to ISO and Responsible Care principles from the 1990's. Both initiatives are voluntary; members of the industry's supply chain globally had been encouraged to adopt these programs and incorporate the principles into their standard operations, to bring about improvements in health, safety and environmental performance and to foster public trust through transparency in communication.

In adapting these kinds of initiatives to accommodate compliance with varied chemical management regulatory requirements, companies had the opportunity to undertake critical review of their entire operation (processes and equipment) and standard operating procedures (SOP's), including all activities in the supply chain. This enabled the opportunities to identify possible areas of redundancies or inadequacies and devise appropriate means of improvement. It also allowed for the possibilities of new discoveries, whether of equipment or device application, processes, intermediate product application or areas for closer inter-department collaboration.

In essence, this exercise forces companies to evaluate what works and what does not; what may benefit from improvement and what may need to be illuminated, from technical operations to business operations. And in the process, they can develop a comprehensive program for fulfilling chemical management regulatory requirements – testing and risk assessment of chemicals, recording keeping of data, reporting required information, filing certification or permit applications – i.e. applying a holistic, sustainable approach.

This approach sustainable approach includes principles of sustainability, which seek to optimize economic benefits to an organization's business, while protecting the environment or eco-system and enhancing human wellbeing. And these are also the ultimate objectives of chemical management regulations, policies and initiatives, to manage chemicals in the supply chain in such a manner that none of these areas are compromised.

BASF, a global industry giant, is one stellar example of a company that employed a holistic approach to chemical management, business operation, environmental, health and safety compliance and ecological preservation. The company developed an analytical model called the SEEBALANCE (illustration below in Fig. 2) which can be used to analyze any of the pillars of sustainability, impacts and cost, and also correlation between any combinations of the three pillars.

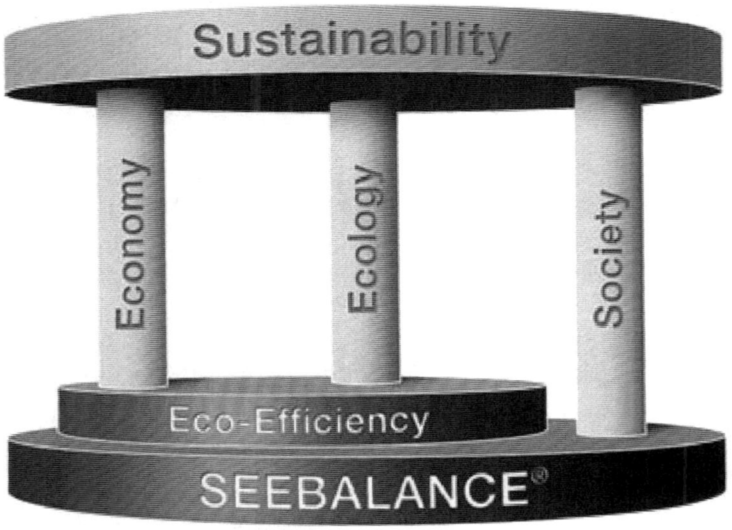

Fig. 2 – Illustration of BASF's Innovative Comprehensive Analytical Tool

Overwhelming public opinion is that regulations are necessary to help attain this tripod balance. And companies are very much aware of the value of meeting public expectation – the result could be the establishment of a lucrative brand or the demise of an unsuccessful one – directly affecting economic performance and the bottom line.

Various institutions and regulatory bodies offer award incentives to companies who attain high levels of achievement in areas such as environmental preservation, health and safety record, product development that enhances ecological preservation (green chemistry) and operational changes that minimizes a company's or chemical's carbon footprint. These incentives also encourage companies to evaluate their overall operation for areas of and opportunities for national and global recognition.

Following are some notable industry achievements and recognition:

- Dow Chemical was honored and a 2012 top 100 innovator by Thompson Reuters, noting "The company connects chemistry and innovation with the principles of sustainability …"*
- Henkel, Serbia and SAFECHEM Europe were two of the 2012 recipients of the Global Chemical Leasing Award – chemical leasing is a service-oriented business model that shifts the focus from increasing sales volume of chemicals towards a value added approach

Chemical companies may also seek to forge synergistic collaborations and strategic alliances as part of their strategy to accomplish chemical management objectives at optimal benefit to business operation. It is from on such collaboration between BASF and various academic research institutes that the SEEBALANCE analytical was developed. Strategic alliances could make major contributions in a product leasing initiative or, more directly, the development of a corporate chemical management compliance system that facilities worldwide can use as a central resource of information on compliance.

Conclusion

The safe management of chemicals has for decades been regarded as an essential policy of operation by all stakeholders in the chemical industry: manufacturers and distributors, intermediate and end product users, the public and regulators. But differences of opinion as to the most viable and balanced way of achieving an outcome acceptable to all stakeholders had persisted until the turn of the 21st century, when a paradigm shift started spreading across the industry. Companies started recognizing the opportunities in developing a comprehensive chemical management program which would address compliance requirements of multiple regulations, while supporting the bottom-line.

A recent survey conducted by *Chemical Watch* indicates that this trend is expected to continue over the next five years, with a growing number of mid-sized and small companies adopting the comprehensive approach. The future promises to be quite sustainable.

Reference Sources

Chemical Watch, 24 January, 2013
www.ICC-Chem.org
*www.dow.com
www.unep.org
www.basf.com

Appendix A

Table 1. Regulations and Policies Containing Chemical Management Components

Administrating Body	Regulations/Policies	Chemical Management Requirements and Standards
United States Environmental Protection Agency (USEPA)	Clean Air Act (CAA)	Risk Management Plan (RMP) – Planned changes to improve safety in chemical handling
	Toxic Substances Control Act (TSCA)	Testing and assessment of chemicals and reporting of data on risks and hazards from use and exposure to them; Certification of information
	TSCA Chemical Data Reporting (CDR)	Periodic update of information on chemicals reported under TSCA;
Occupational Safety and Health (OSHA)	Process Safety Management (PSM) of highly hazardous chemicals	Comprehensive management program for preventing / mitigating hazards associated with processes using highly hazardous chemicals;
European Chemical Agency (ECHA)	Registration, Evaluation, Authorization of Chemical Substances (REACH)	Testing of chemicals and assessment of data to determine risk; Develop and recommend appropriate risk management measures; Register
Inter-organizational Program for the Sound Management of Chemicals (IOMC)	Global Harmonized System (GHS) of chemical classification	Internationally accepted system for the classification and labeling of chemicals
International Conference on Chemical Management (ICCM)	Strategic Approach to International Chemicals Management (SAICM)	Framework to develop and implement sound chemical management practices globally

Risk-Based Approach to Building Designs and Decision Making

John N. Dyer P.E., Anay P. Raibagkar P.E., Anthony G. Sarrack (Baker Engineering and Risk Consultants, Inc.)

Introduction

Accidental releases of flammable and toxic materials from hazard sources of different sizes ranging from small leaks to full-bore pipe ruptures can represent a wide range of impacts to buildings. Facility siting for petrochemical and chemical plants is required to cover credible accident scenarios. Results of such studies often include catastrophic consequences that may be impractical to eliminate through structural designs. Rather than designing buildings for a high level of protection against the most severe credible events, a risk-based approach provides a practical and a cost effective approach to maximize safety.

Two fundamental principles are required to be met for a design to qualify as "Risk-Based". One principle is that each occupant of the building has a minimum level of protection. This principle is achieved by designing the building in such a way that each occupant's individual risk is within defined tolerance criteria. The other principle is that the building is designed in a way such that it reduces the overall societal risk to a level that is as low as practical. This principle is achieved by comparing costs of enhanced designs against the safety benefit they afford and implementing as much mitigation as is practical. Employing these fundamental principles ensures that cost is optimized for a given risk level, allowing risk mitigation resources to be optimized.

This paper outlines a methodology to perform a risk-based building design, accounting for predicted impacts of a wide range of credible accidents, associated frequencies, and building occupancy. The necessary input data for this type of design is calculated in a Quantitative Risk Assessment. The proposed methodology will help designers establish practical structural requirements to minimize risk while achieving cost effective designs.

Building Blast Damage and Occupant Vulnerability

The expanding pressure wave resulting from an accidental explosion will load objects in its path with a relatively uniform pressure applied to all exposed surfaces, as illustrated in Figure 1. The sides of a structure facing the source of the explosion will cause the wave to reflect, increasing the magnitude of the overpressure on those surfaces.

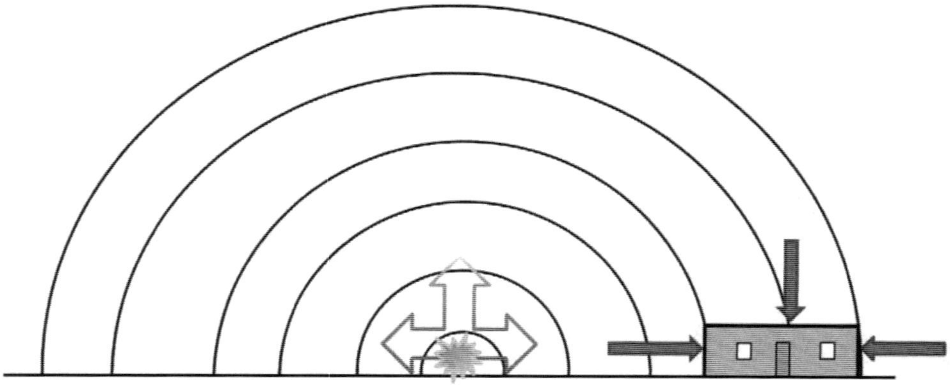

Figure 1. Pressure Wave Loading of Building

The explosion pressure may be considered to be applied instantaneously and to decay to the ambient pressure linearly, as illustrated in Figure 2. The time that the overpressure is applied on a building's surfaces is usually a small fraction of a second. The impulse of the blast load is defined as the integral of the load function. For the idealized triangular load history illustrated in Figure 2, the impulse may be approximated as the area of the triangle. The intensity of a blast load on a building surface may be expressed in terms of its peak pressure and duration or by its peak pressure and impulse.

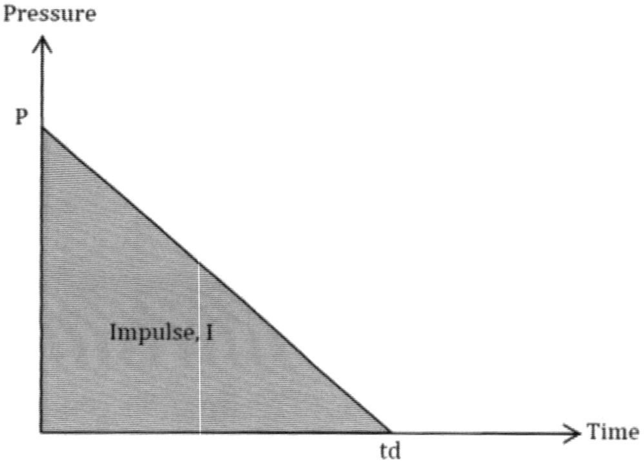

Figure 2. Idealized Overpressure History

The blast will cause structural elements on the loaded surfaces of a building to respond in an oscillatory fashion. Components that are not catastrophically damaged, will reach a peak deflection followed by a rebound phase. The oscillations will become smaller with time until they stop. The level of protection that a building affords its occupants is typically dependent on its ability to resist blast loads without sustaining significant damage. Therefore, estimating the building damage level (BDL) caused by explosions is important.

Occupant vulnerability (OV) is defined as the fraction of the building population who will sustain fatal injuries. Since the OV is connected to the damage sustained by the building, establishing a BDL categorization is useful to assess building damage and occupant vulnerability. Table 1 provides a summary of typical building damage levels with a brief description of structural component damage.

Table 1. Building Damage Level Categories [1]

BDL	Damage Description
1	Minor damage to roof and walls
2	Moderate or lower damage to roof and load-bearing components
2.5	Localized Major damage to roof or walls. Moderate damage to other components
3	Major damage to load-bearing elements or roof
4	Failure of load-bearing elements or roof collapse

The two lesser damage levels (BDL 1 and BDL 2) are typically associated with minor to moderate component damage particularly to the sides of the building facing the explosion, as illustrated in Figure 3. BDL 1 is used in instances where very low damage is sustained, and BDL 2 corresponds to cases in which some building components may have sustained moderate damage. When a building sustains BDL 1 or BDL 2, personnel inside the building are not considered to suffer life-threatening injuries.

Figure 3. Minor Building Damage

BDL 2.5 is used to represent instances where major building damage is sustained including some structural failures on the walls. BDL 2.5 is associated with low (not negligible) occupant vulnerability, since life-threatening injuries could typically only result to personnel located near the most heavily damage areas, as those shown in Figure 4.

BDL 3 and BDL 4 are associated with widespread structural damage that could potentially compromise the integrity of the entire building. Occupant vulnerability is typically significant [1] at BDL 3 and 4. In particular, BDL 4 is used to represent collapse of the roof, which can result in a large portion of the building population sustaining fatal injuries ($0.5 < OV < 1.0$).

Figure 4. Major Building Damage

Building Blast Assessment

Designing buildings to resist explosion blast loads requires structural methods that allow quantifying the level of building damage and estimating occupant vulnerability. Since the severity of a blast load cannot be fully assessed without knowing its peak pressure and impulse, the structural method should take into account not only the pressure but also its duration. Dynamic structural analyses will allow taking into account the duration of the blast load. In order to estimate structural damage, the elastic-plastic behavior of the structure must be modeled. Since more than one explosion case is often necessary to be evaluated, the structural methodology should preferably be based on simplified analyses to avoid expensive structural calculations. The Single-Degree-of-Freedom (SDOF) methodology allows engineers to perform relatively inexpensive calculations with a reasonable level of precision. The SDOF methodology can also be used to construct component damage curves based on pressure-impulse (P-i) combinations that result in equal component response. The component damage curves can be combined to construct building damage Pressure-impulse diagrams as the one shown in Figure 5. The SDOF methodology is a widely accepted analysis technique and is described in several references [2, 3].

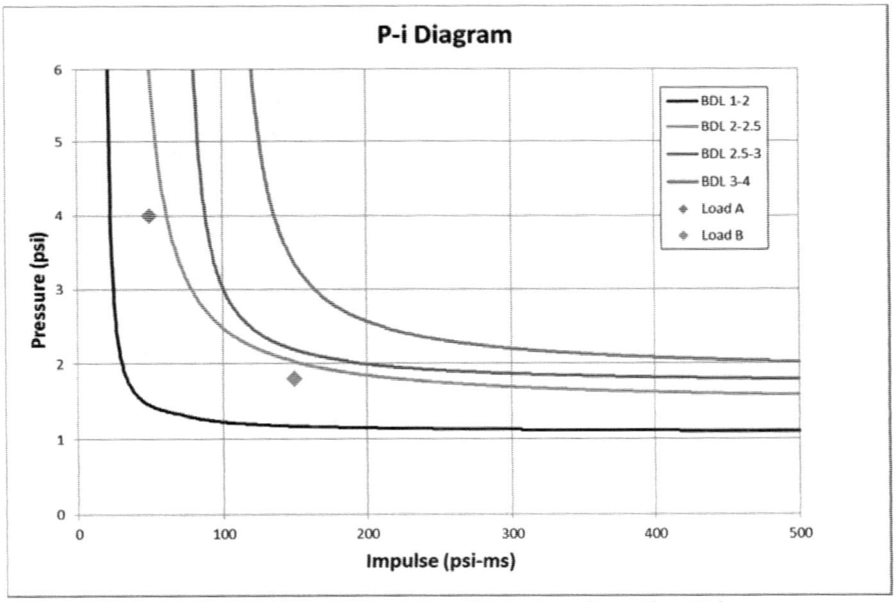

Figure 5. Building Damage Pressure-impulse Diagram

The P-i curves represent boundaries of BDL regions. Blast loads can be plotted on the P-i diagram to perform graphical blast assessments. Since the curves represent blast loads that result in a specific damage, a blast load represented by a P-i point will result in higher or lower damage than that represented by the curve depending on their relative position. For example, the two blast loads with different pressure and impulse shown in Figure 5 would cause the same damage level since they are bound by the same BDL curves. P-i diagrams are particularly useful when damage from multiple blast loads need to be assessed, since evaluating multiple component under numerous blast loads can be costly.

Occupant Vulnerability

Estimating occupant vulnerability (OV) is the quantification of loss of human life inside a building from an explosion. OV multiplied by the number of people in a building is typically the "consequence" used in calculating explosion risk. The main driver of OV is BDL [1] and can be estimated by calculating damage on the roof and walls, as illustrated in Figure 6.

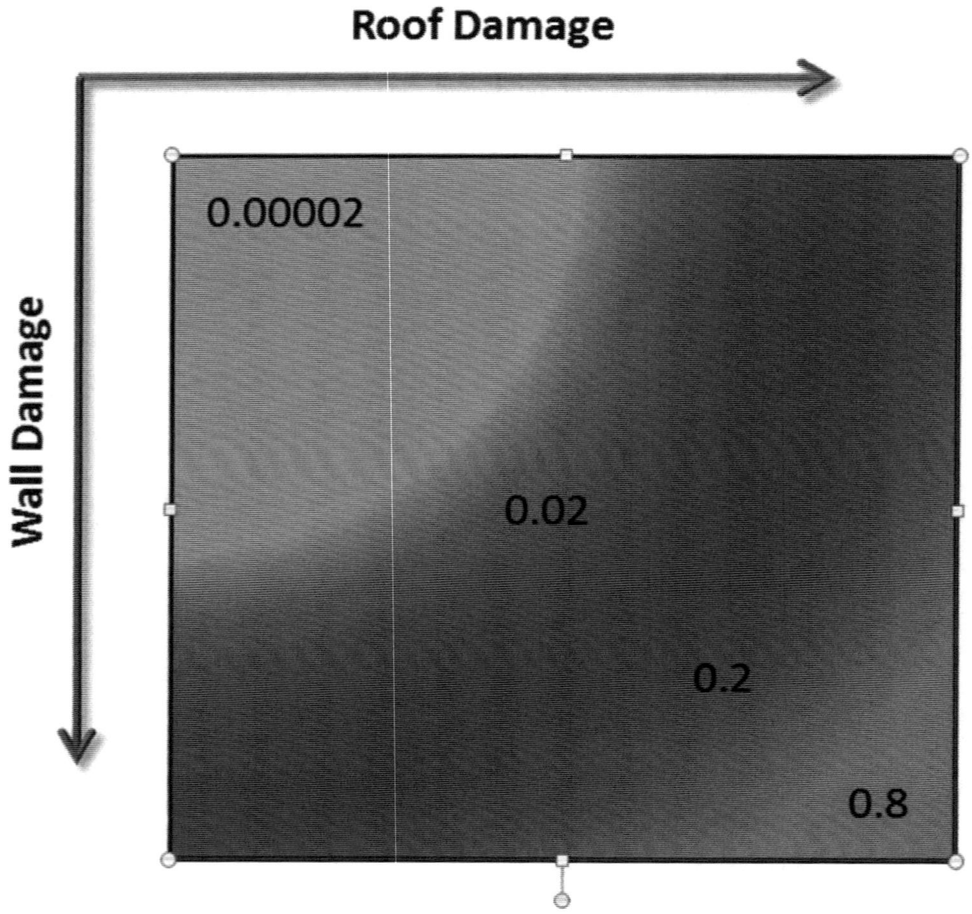

Figure 6. OV as a Function of Structural Damage (typical)[1]

However, factors such as the distribution of the population inside the building, the amount of window coverage on exterior walls, the structural system of the building, and the weight of the construction can influence the OV [1].

[1] This is a generic figure. The actual values of OV will depend on several factors such as type of construction, window coverage on walls, perimeter population in the building etc. A detailed methodology for calculating OV is given in [1].

Table 2. Typical Occupant Vulnerability Values (for Each BDL)

BDL	Typical Occupant Vulnerability	
	Steel Frame with Metal siding	**Concrete Frame with masonry walls**
1	0.00002	0.00001
2	0.00008	0.00005
2.5	0.015	0.02
3	0.2	0.3
4	0.5	0.95

The consequences in terms of fatalities can be calculated for a building for each explosion scenario by assessing building damage and the corresponding occupant vulnerability which is multiplied by the occupancy of the building to estimate the total number of fatalities. Therefore, the consequence of an explosion (for a particular scenario) can be calculated as

$$\text{Consequence}_{\text{bldg-exp}} = \text{OV}_{\text{bldg-exp}} \text{ x } \text{Pop}_{\text{bldg.}}$$

Where,

$\text{OV}_{\text{bldg-exp}}$ – Occupant Vulnerability corresponding to the BDL

$\text{Pop}_{\text{bldg.}}$ – Average occupancy in the building

Risk Assessment[2]

Risk is defined as the product of consequence (fatalities / event) and frequency (events / year) and is presented in terms of fatalities per year. Therefore, risk (for a particular scenario) can be calculated as

$$\text{Risk}_{\text{scenario}} = \text{F}_{\text{scenario}} \text{ x } \text{Consequence}_{\text{scenario}}$$

The frequency of each explosion is calculated by multiplying the release frequency by the applicable conditional probabilities of factors affecting the likelihood of the explosion such as

- Equipment failure rates
- Weather conditions
- Wind direction
- Ignition probability, and
- Time slot probability

Therefore, the frequency of the explosion scenario can be calculated as

[2] The risk assessment method given in this paper is for explosion risk only. The method can be extended to other hazards such as toxic and fires.

$$F_{scenario} = R_{eqpt} \times C_{eqpt} \times P_{weather} \times P_{wind} \times P_{ignition} \times P_{time}$$

When blast loads are relatively low at the location of interest, designing a building to sustain low or moderate damage is typically enough to ensure that the OV is negligible, and risk will be typically low. However, when a building is potentially subjected to high overpressures, as illustrated in Figure 7, the structural requirements to mitigate higher than moderate (BDL 2) damage can be excessive making the building design cost prohibitive.

In order to achieve a cost-effective design and provide an adequate level of protection under relatively high blast loads, the design of the building must take the frequency of the explosion scenarios and be based on achieving a tolerable risk level. Thus, the total societal explosion risk for a building can be calculated by adding risk from all postulated scenarios as shown below.

$$\text{Societal Explosion Risk} = \sum \text{Risk}_{scenario} = \sum (F_{scenario} \times \text{Consequence}_{scenario})$$

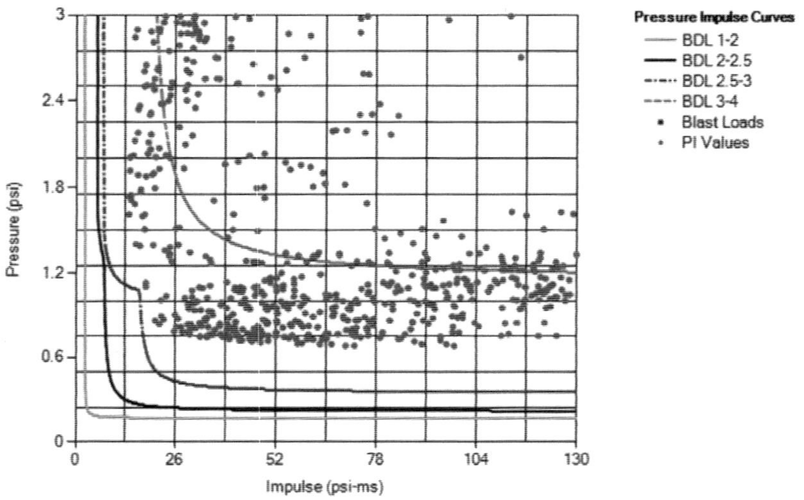

Figure 7. P-i Diagram with All Credible Events

Risk-Based Structural Design

The explosion risk methodology summarized in this paper can be effectively used in selecting a cost effective design as indicated in Figure 8. The figures show explosion risk vs. additional building costs for four design options. The 'Original' design can be considered as a baseline (cheapest design that will satisfy the local building codes only) design with no extra resources spent for the purposes of explosion risk mitigation. The remaining three options, give considerably lower risk compared to the 'Original' baseline design. However, the chart on the right indicates that additional resources will be required to lower the risk below the tolerable level. As indicated in the 'Explosion Risk' chart, the risk associated with the 'BRM' option is just below the tolerable limit of 8.0E-04 and the corresponding cost of $360,000 is the lowest among the three options under the tolerable risk. Thus, in this particular example, the BRM option can be considered as the most effective risk mitigation solution and can be selected for construction.

However, this choice is based on explosion risk only. Similarly, risk from toxic releases and fire hazards may be present and the BRM option may not be a suitable choice. Thus, a similar risk based method should be applied in selecting a cost effective design option and making decision for effectively allocating resources and maximize safety.

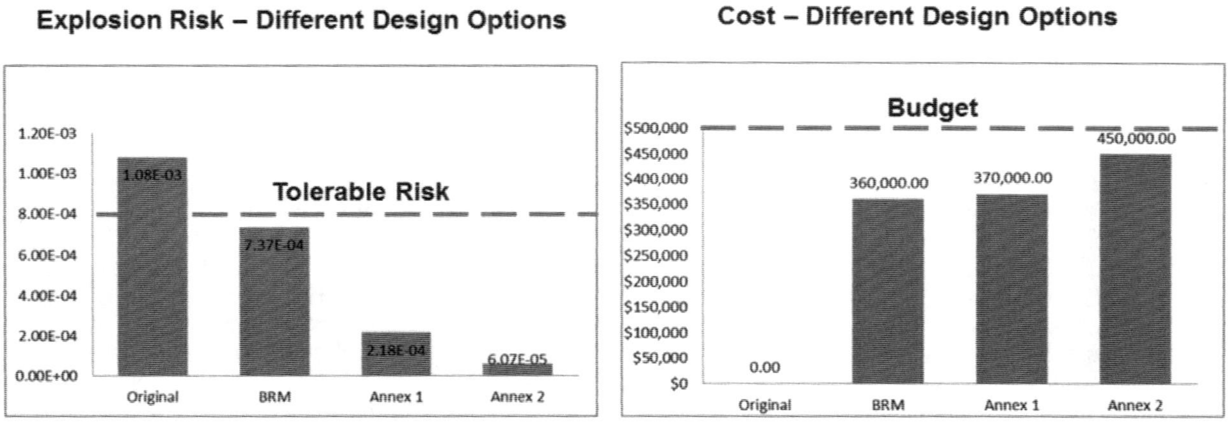

Figure 8 Comparison between cost of a building vs. Risk to building occupants

References

1. Oswald, Charles J. and Quentin A. Baker (1999), "Vulnerability Model for Occupants of Blast Damaged Buildings," *Prepared for Presentation at the 34th Annual Loss Prevention Symposium, March 6-8, 2000, Session 3.*
2. Biggs, J.D (1964), "Introduction to Structural Dynamics," McGraw-Hill Publishing Company, New York.
3. ASCE, ASCE Task Committee on Blast Resistant Design, Design of Blast Resistant Buildings in Petrochemical Facilities, American Society of Civil Engineers, 2nd Edition, New York, 2010.

(32c) EPA's Greenhouse Gas Reporting Program: Lessons Learned for Onshore Petroleum and Natural Gas Production Operations

Karin Ritter, American Petroleum Institute, Washington DC, DC, Miriam Lev-On, The LEVON Group, Thousand Oaks, CA and **Theresa Shires**, URS Corporation, Austin, TX

The Environmental Protection Agency (EPA) mandatory greenhouse gas (GHG) reporting program applies to facilities and operations whose annual GHG emissions exceed 25,000 tonnes of carbon dioxide equivalent. For the Petroleum and Natural Gas Systems sector, 40 CFR Part 98 Subpart W lists the requirements for quantification and reporting of GHG emissions. The methodologies specified for the eight industry segments within the sector, range from application of specified emission factors, engineering calculations, and direct emissions measurements. Notably, the Onshore Petroleum and Natural Gas Production segment is comprised of a large number of emission sources within "facilities" that cover multiple operations within a single hydrocarbon basin. These requirements necessitate installation of proper instrumentation, robust data collection practices, and Information Technology (IT) systems that enable a smooth interface with EPA's electronic greenhouse gas reporting tool (eGGRT).

This paper will address the lessons learned from the initial phases of reporting for the Petroleum and Natural Gas systems and will emphasize the unique aspects of reporting GHG emissions for onshore oil and natural gas production operations. The discussion will address the need to balance reliable information and reporting burden and will highlight industry's commitment to continuous improvement. It will address how reporting provides companies a means to highlight altering practices like venting and flaring minimization, and reduction of environmental impacts of operations. Such systematic data collection, as mandated by the EPA GHG reporting program, may not only help target areas for future reductions but also demonstrate significant voluntary progress toward continually 'greener' operations.

(32d) Trials and Tribulations in Carbon Foot Printing

Dr. Sunil Hangal, Environmental Compliance Assurance Management (ECAM), Bedminster, NJ

Carbon foot printing is a determination of the total greenhouse gas emissions from an organization, product or event. This presentation describes the trials and tribulations in obtaining accurate and verifiable total greenhouse gas emissions through the complete lifecycle of a product. Complete lifecycle emissions include all direct and indirect greenhouse gas emissions including carbon dioxide, nitrous oxide, and hydrofluorcarbons (HFCs) through the raw material to disposal after end use of a product.

In this study, greenhouse gas emissions were estimated for a textile fabric sold wholesale to customers. This company identified two popular products as a test case to obtain carbon footprint data. These two products were high end product with higher recycled fiber content and a product line with 100 % natural fiber origination in the US. The product lifecycle boundaries were agricultural emissions, manufacturing, transportation and storage and use of product. The goal was to obtain carbon dioxide equivalent emissions per unit measure of cloth that could be provided to end use customers.

Detail lifecycle planning was done and emission was estimated using known emission factors and activity data across the lifecycle. There were several issues that affected the accuracy of data such as variability of fertilizer use in growing cotton, reliability of some emission factors, and apportionment of activity for transportation sources, amongst others. This presentation will also discuss practical issues in working with suppliers and vendors down the supply chain in order to obtain accurate and verifiable activity data. In addition data rollout planning and difficulties with benchmarking will be discussed.

(45a) Pilot and Full-Scale Application of Technologies for Mercury Control

Noah Meeks, Research & Environmental Affairs, Southern Company Services, Inc., Birmingham, AL

With the final MATS rule going into effect in 2015, electric utilities must now consider mercury control as a needed technology in their emissions train. This talk wiill cover recent full-scale injections of activated carbon and alkali injection, as well as pilot-scale studies of non-halogenated carbons or non-carbon sorbents. We will also discuss recent mixed matrix membrane pilot studies. **(Most recent data will be included in the abstract closer to the meeting)

(45b) Acid Tunneling Vs Acidizing : A Case Study of Rudatain Oilfield-Kuwait

Adel Elsharkawy, Petroleum Engineering, Kuwait University, Kuwait, Kuwait and Mohamed Al-Hadher, Pet Eng, Kuwait Oil Company, Safat, Kuwait

Acid Tunneling Vs Acidizing : A Case Study of Rudatain Oilfield-Kuwait

By Mohammed Al-Hadher and Adel Elsharkawy

Kuwait University, College of Engineering & Petroleum

Po Box 5969 Safat 13060 Kuwait

Abstract

Acidizing is one of the most important techniques used in oil industry in order to enhance the reservoirs and increase oil production. There are three types of acidizing techniques which are matrix acid stimulation, acid fracturing, and acid tunneling. In this paper a comparison these types of acidizing is discussed. The case study of RA-0007 acid tunneling completed in Kuwait Oil Company is presented. It represents the first dendritic with extensive worm holes created by acid tunneling for maximum reservoir contact. The whole effect is aptly demonstrated by the change in PI of the well before and after the acid tunneling as mentioned before. Analysis of the well designed and implemented job showed that two tunnels were created successfully in MA-C formation through a two days rig-less operation. The results indicated an improvement in the well productivity index (PI) which had been increased from 1.3 to 3.87 bpd/Psi.

(45e) Sustained Casing Pressure Modeling and Well Integrity

Tony Rocha-Valadez, Mary Kay O'Connor Process Safety Center, Artie McFerrin Department of Chemical Engineering, Texas A&M University, College Station, TX, A. Rashid Hasan, Harold Vance Department of Petroleum Engineering, Texas A&M University, College Station, TX, C. Shah Kabir, EPT, Hess Corporation, Houston, TX and Sam M. Mannan, Artie McFerrin Department of Chemical Engineering, Mary Kay O'Connor Process Safety Center, Texas A&M University, College Station, TX

SUSTAINED CASING PRESSURE MODELING AND WELL INTEGRITY

T. Rocha-Valadez[1*], A.R. Hasan[2], C.S. Kabir [3] and S.M. Mannan[1]

*[1]Mary Kay O'Connor Process Safety Center, Artie McFerrin Department of Chemical Engineering, Texas A&M University, College Station, Texas 77843-3122, US., email: *tonyrochav@tamu.edu, mannan@tamu.edu.*

[2] Harold Vance Department of Petroleum Engineering, Texas A&M University, College Station, Texas 77843-3122, US, email: rashid.hasan@pe.tamu.edu.

[3]Hess Corporation, 1501 McKinney Street, Houston, TX 77010, US, email: skabir@hess.com.

With an increasing demand for hydrocarbons and other sources of energy, the methods for delivering such energy needs have become complex. In the case of the oil and gas industry, exploration and production have gone to deeper waters and deeper horizons than ever before, resulting in drilling several miles into the earth's core. Deeper well penetration means encountering increasingly high formation pressures and temperatures. For example, in deepwater drilling, well construction and production management deal with significant heat transfer between the produced fluids and its surroundings. The high-rate production and the attendant heat transfer demand close attention in terms of operational safety and potential environmental impact. One phenomenon that occurs in producing wells is sustained casing pressure (SCP), which presents a risk to both process safety and the environment.

In oilfield terminology, the casing is a steel pipe cemented to the surrounding formation. Multiple concentric casings are involved in a typical well construction. The casing immediately surrounding the production tubing is known as the production casing, which is most vulnerable to thermal and mechanical stresses and fluid leakage. The cement sheaths provide zonal isolation, support the casing, and protect it from corrosive fluids. Ideally, pressures monitored at the wellhead for multiple casings should change minimally after being bled off through a needle valve at the surface. When the pressure rebuilds after being bled off, that casing is said to exhibit SCP [1].

SCP occurs because of stresses that the system experiences owing to fluids production up the tubing. Besides the normal source of SCP, a poor cement job may allow hydrocarbon migration from the producing or other horizons and accumulation into one or more of the casings. The risk of not attaining well integrity can range from the activation of rupture discs to the blowout of hydrocarbons. Because of the safety and environmental dangers of SCP, regulations require remedial treatments for a casing string that exhibits significant SCP problems. Unfortunately, available methodologies to test for SCP are often qualitative and empirical in nature [2, 3], rendering them inapplicable in many situations. No practical systematic testing procedure has emerged to quantify SCP.

In this paper, we present a model for SCP that is rooted in the transport processes of the system. The model assumes gas leaking through the cement barrier, and then migrating up the annular-liquid column. It accounts for the fluid-temperature profile along the wellbore and liquid compressibility. Some robust assumptions about the transport processes allow us to write the governing equation as a first-order, linear-differential equation that is solved with appropriate boundary conditions. The resulting algebraic expression for the casing pressure rise as a function of time paves the way for all applications. A comparative study of the model's performance with field data suggests excellent agreement for wells where the casing pressure increase is owes gas influx into the annulus. The proposed model could form the basis for standardization of SCP testing. This study also offers some test procedures to assess SCP-related safety concerns.

References

1. Bourgoyne Jr., A.T.J., S.L. Scott, and W. Manowski, ***A Review of Sustained Casing Pressure(SCP) Ocurring on the OCS***, 2000, Final Report Submiteted to MMS.

2. Xu, R. and A.K. Wojtanowicz, ***Diagnosis of Sustained Casing Pressure from Bleed-off/Buildup Testing Patterns.*** SPE, 2001: p. Productions and Operations Symposium.

3. Wojtanowicz, A.K., S. Nishikawa, and R. Xu, ***Diagnosis and Remediation of Sustained Casing Pressure in Wells***, July 2001, MMS.

(45f) Evaluation of Particle Growth in an Electrochemical Fluidised Bed Reactor

Emmanuel Ehirim, Rivers State University of Science and Technology, Port Harcourt, Nigeria

The process of electrodeposition within electrochemical reactor results in the increase of particle sizes, which increases the volume of the bed. This work entails a mathematical model to describe the effect of this phenomenon on the behaviour of a three-dimensional fluidised bed electrode operating under the condition of electron transfer. The model incorporates the pseudo-stationary behaviour in the reactor since the bed increases in volume due to the particle size growth. The Orthogonal Collocation method was used to solve the developed differential equations. The overpotential distribution and the reaction rate within the reactor have been calculated. The principal results are the effects of the electron transfer operation on the superficial displacement with time. The effects of total applied current, electrode geometry and the bed porosity on the behaviour of the system have been analysed. The result shows a very gradual increase in the bed height.

(45g) Debottlenecking Gas Dryers and Desiccant Adsorbers

Benjamin A. Schmitt, Hydrocarbon Processing Industry (HPI) - Business Unit, Johnson Screens, Inc., New Brighton, MN and Michael Ekholm, Engineering, Johnson Screens, Inc., New Brighton, MN

Debottlenecking Gas Dryers and Desiccant Adsorbers

Though Maximization of the Useable Bed Volume with Johnson Screens' SSG™

Benjamin Schmitt, PE

Senior Mechanical Engineer, Technology

Michael Ekholm, PE

Applications and Product Design Manager

Johnson Screens Inc. St. Paul, MN

AIChE Spring Meeting, April 2013

13th Topical Conference

San Antonio, TX

Introduction:

The following presentation of the Shaped Support Grid (SSG™) releases additional supporting data and updates to the initial presentation given by Robert G. Norell at the 2010 AIChE Spring Conference.

Limitations on ethylene dryers and desiccant adsorbers can create bottlenecks in the production process which can lead to a need for increased dryer and adsorber capacity. Capital intensive investments may be necessary in order to eliminate the bottleneck which may include the addition of an additional vessel. With limits on the plot plan, this may not be a workable solution. Adding to the problem is the amount of time consumed in engineering, planning and construction, which can have negative effects on the operation of the ethylene unit. Increased desiccant volume and operational efficiency can increase the unit's production with minimal capital expenditures. Johnson Screens', patent pending, SSG™ can add the increased process efficiency and cycle life necessary to extend the operating capability of the unit.

Extending the operation, though debottlenecking, can be achieved by maximizing the time before breakthrough of the media bed. There are three primary functions that maximize the breakthrough time:

Proper gas distribution in both normal operation and regeneration conditions which maximizes the bed utilization.

Increased molecular sieve or desiccant bed volume are achieved through the elimination of traditional bed supports comprised of support grids and beams. The support interface for the screen configuration also allows for thermal expansion typically present in vessels with a thermal cycle as part of the process. The design allows for thermal cycling without the risks of seal or structural failure inherent in other solutions.

Minimizing the bed volume consumed by the distribution and support system increases the net desiccant volume lost when utilizing alternative bed supports.

Johnson Screens ***SSG*** is designed to provide the industry with an alternative, cost effective, approach to increasing capacity. Supporting data and preliminary case studies will be depicted to quantify the benefits that are achievable through the implementation of the ***SSG***.

Conclusions:

The use of a Johnson Screens ***SSG*** provides a reliable immediate solution to reduce capacity bottlenecks. The ***SSG*** increases the bed volume, when compared to traditional bed support arrangements comprised of grids and beams. The increase in desiccant volume will allow for increases in feed flow rate and/or adsorbtion cycle time. The ***SSG*** maximizes the flow distribution during desiccant regeneration, maximizing the utilization of the complete cross section of the vessel. This solution can be implemented in both new and existing vessels, providing consistency in design between planned increases in performance of current vessels and future expansions.

The modular design of the ***SSG*** allows for easy installation and maintenance. Connection of the ***SSG*** to the vessel is performed within the outlet nozzle, eliminating all welding on the pressure vessel and eliminating any need for recertification of the pressure vessel after installation. The design allows for ease of dismantling for inspection of the vessel head during routine maintenance and inspections. The low profile will also reduce interference with any required media exchanges. The design package of the ***SSG*** provides the end user with a holistic solution that fulfills debottlenecking needs with limited capital expenditures.

Biography of Speaker:

Benjamin Schmitt is a Senior Mechanical Engineer in Johnson Screens' Hydrocarbon Processing Industry Technology Group. Ben graduated from North Dakota State University in 2004 with a Bachelor of Science degree in Mechanical Engineering. He currently working toward his Master's in Business Administration from Hamline University. Ben is member of AIChE and is a registered professional engineer in the state of Minnesota.

AMACS
PROCESS TOWER INTERNALS

AIChE 2013 Spring Meeting Poster board

Design Considerations for effective liquid removal In Suction Drums

Presented By:

Kanti Patel, PE

Compressor down again?

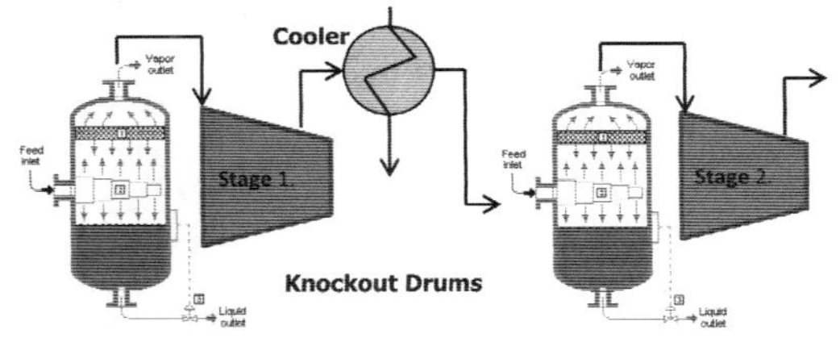

Is liquid carryover the culprit?

<u>We have heard it all before</u>:

-Premature compressor shutdown.

-Production time lost.

-High yearly compressor maintenance costs.

-Impeller damage due to pitting corrosion.

-Corrosion due to salt deposits.

AMACS
PROCESS TOWER INTERNALS

Common Design Problems with Suction Drums which promotes liquid carryover

1. Choosing the correct mist eliminator (M.E. equipment)
2. Knowing where to install the M.E. in a suction drum.
3. Consider a high mist loading in the gas when selecting your M.E.
4. Fouling considerations.
5. M.E. disengaging space, bad gas distribution & liquid slug in the feed.
6. One micron and great droplets, versus submicron mist removal.
7. High viscosity liquids.
8. Dislocated M.E. due to pressure surges.
9. Through put exceeds original design capacity.
10. Capacity range too large for suction drum.

AMACS
PROCESS TOWER INTERNALS

Choosing the correct Mist eliminator equipment

- Proper sizing

+ Use Souders-Brown equation to find optimum velocity. $$\mathbf{Vd = K} \cdot \sqrt{\frac{Pl}{Pg} - 1}$$
+ Select the correct vapor velocity factor K. based on the operating pressure, and type of equipment.
  - Lower K-factor 0.01 for every 100 psi
  - K factor to select when using:

Mesh Pad	K= 0.1 - 0.55
Chevron Vane	K= 0.45 - 0.65
MultiPocket® Vane	K= 0.75 – 1.0

AMACS
PROCESS TOWER INTERNALS

Choosing the correct Mist eliminator equipment cont.

- Selection based on droplet size.

Table 2. Droplet sizes (water in air) typically captured with 99.9% efficiency by mist eliminator elements of various types

Element type	Size range microns
Fiber candles or panels	0.1 and larger
Mesh with co-knit yarn	2.0 μm and larger
0.006-inch knitted mesh	5.0 μm and larger
0.011-inch knitted mesh	10 μm and larger
AMACS MultiPocket ® Vane	10 μm and larger
Conventional vane arrays	15 μm and larger

AMACS
PROCESS TOWER INTERNALS

Where to install the M.E. in the suction drum

- Improperly installed M.E. can cause severe liquid carryover problems.
- General accepted spacing guidelines.

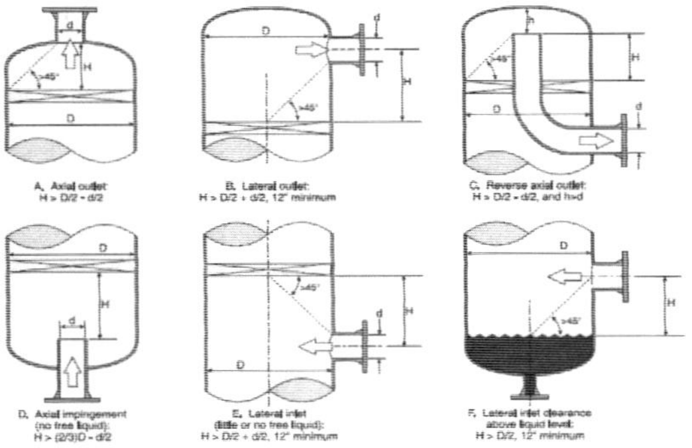

AMACS
PROCESS TOWER INTERNALS

Consider high mist loading when selecting your M.E. equipment

+ High mist loading in the gas stream can flood your M.E. equipment and cause liquid carryover.

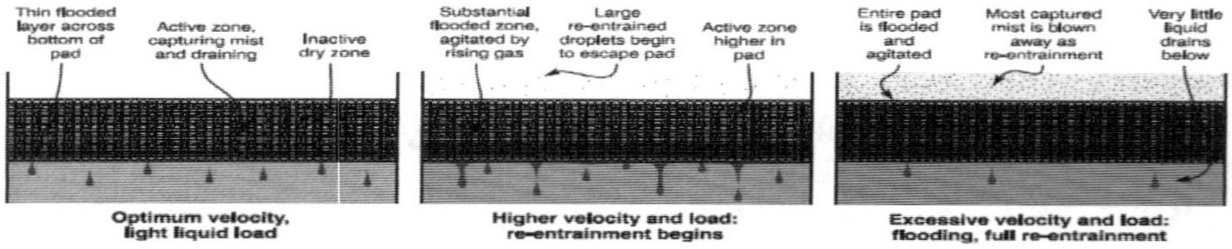

Figure 4. Envisioning mesh pad performance degradation as liquid and gas loads increase (vertical cross-sections)

AMACS
PROCESS TOWER INTERNALS

Consider high mist loading when selecting your M.E. equipment cont.

+ Limited drainage capabilities
  - mesh M.E. <0.5 gal/FT2/min
  - Vane type M.E. <2 gal/FT2/min

+ Select a combination of vane & mesh mist eliminators

AMACS
PROCESS TOWER INTERNALS

Fouling Applications

+ Fouling of mesh mist eliminator will cause liquid carryover.

+ For mild fouling install a water spray system below the mist eliminator for occasional inline cleaning (based on Delta-P)

+ For consistent fouling problems, install a Vane mist eliminator upstream of the mesh mist eliminator.
  - Vanes remove solids and bulk liquids therefore, protecting the downstream mesh mist eliminator

AMACS
PROCESS TOWER INTERNALS

Fouling Applications cont.

+ For consistent fouling problems where the solids are "sticky" install a water spray system below the vane mist eliminator.

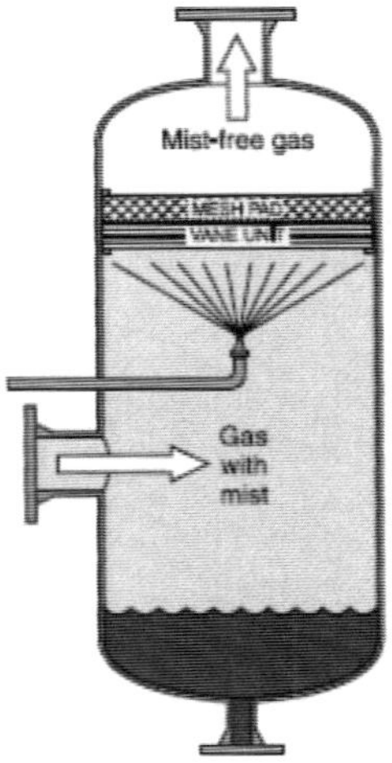

AMACS
PROCESS TOWER INTERNALS

Bad Gas distribution, limited disengaging space and liquid slugs.

+ Bad gas distribution, limited disengaging space in a suction drum can cause severe liquid carryover.

+ CFD analysis can assist in locating high velocity areas.

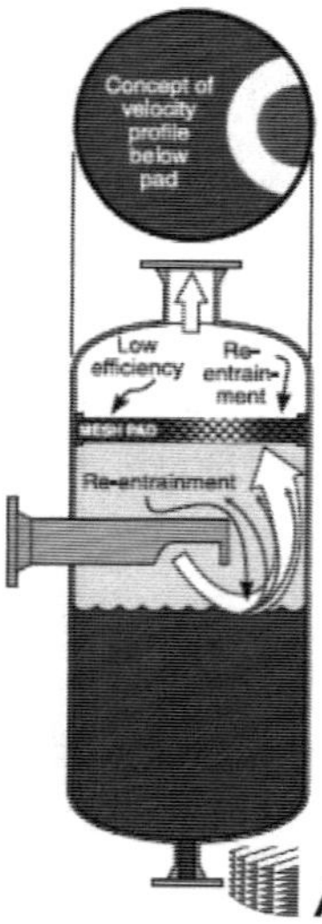

AMACS
PROCESS TOWER INTERNALS

Bad Gas distribution, limited disengaging space and liquid slugs cont.

+ Use inlet diffuser to force a uniform gas distribution in the suction drum.

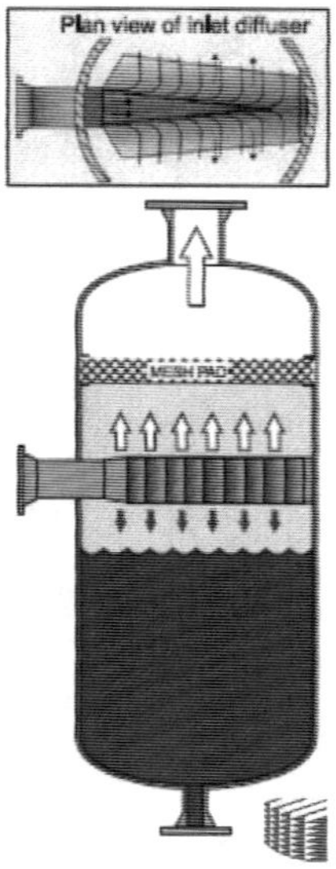

AMACS
PROCESS TOWER INTERNALS

Inlet Feed Device

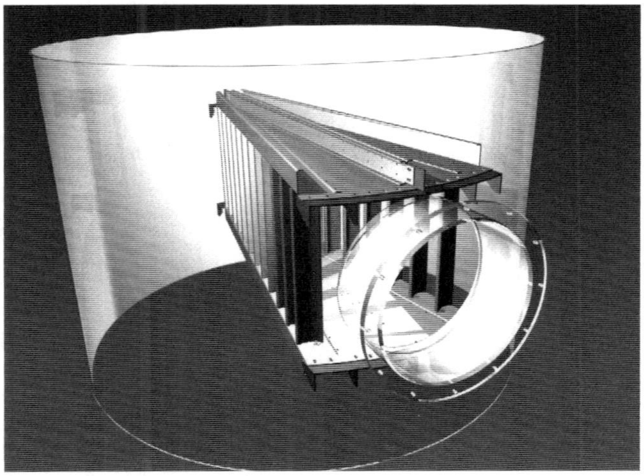

AMACS
PROCESS TOWER INTERNALS

Inlet Feed Device

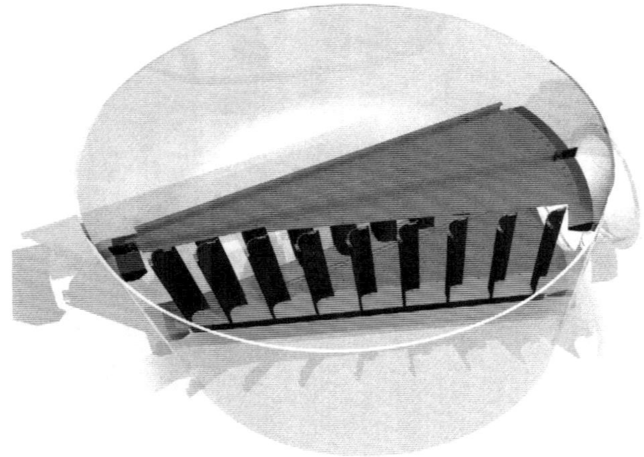

AMACS
PROCESS TOWER INTERNALS

Bad Gas distribution, limited disengaging space and liquid slugs cont.

+ Use a perforated distribution plate downstream of the M.E. for additional distribution assistance.

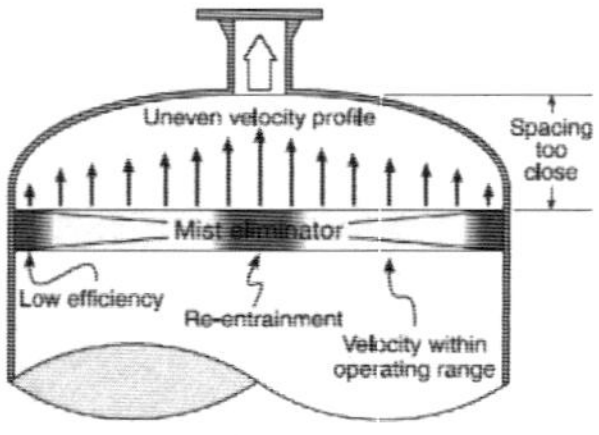

Figure 7. Example of mist eliminator performance degradation due to uneven velocity profile

+ Inlet diffusers also handle incoming liquid slug which otherwise can cause flooding of the Mist eliminator.

AMACS
PROCESS TOWER INTERNALS

Know what mist size droplets You are dealing with

+ Know if your mist is > 1 micron or submicron.
+ When gas travels from A to B, condensation in the pipeline will cause a fine mist (fog or submicron droplets)

  Examples:

  - Incoming gas from an offsite location
  - Fuel gas to furnaces/heaters with low NOx burners

AMACS
PROCESS TOWER INTERNALS

Know what mist size droplets You are dealing with cont.

+ Submicron mist is not captured by conventional M.E.

+ Submicron mist requires a coalescing filter, typically
installed in a two stage filter separator.

AMACS
PROCESS TOWER INTERNALS

Know what mist size droplets You are dealing with cont.

<u>1^{c} stage;</u> submicron droplets are coalesced to > 10 micron droplets.

<u>2^{c} stage;</u> greater than > 10 micron droplets are ejected from the coalescing filter and captured by a vane, which captures droplets > 8 micron.

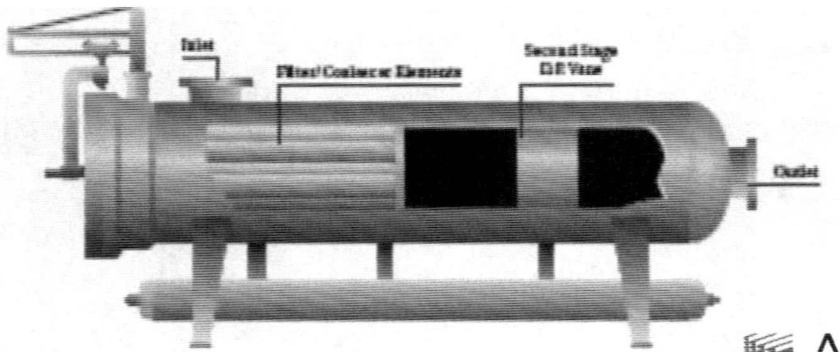

AMACS
PROCESS TOWER INTERNALS

High viscosity liquid as mist

+ Viscous liquids drain slower out of the M.E. and could cause localized flooding.
+ Use a mesh mist eliminator with an open drainage layer on the bottom to improve liquid drainage.
+ Use a lower K- value (lower face viscosity) for the M.E. To allow additional drainage time.
+ Use a vane type M.E. upstream of the mesh mist eliminator, to remove the bulk of the viscous liquid, so the M.E. can effectively handle the residual liquids.

Dislocated M.E. due to System pressure surges

+ M.E. sections can be dislodged into the downstream equipment, which then creates massive carryover.
+ Standard M.E. equipment often gets installed and secured to the support ring with tie wire.

Dislocated M.E. due to System pressure surges cont.

+ Where system pressure surges are common, there are more secure ways to install the M.E.

  - dual support rings
  - through bolts
  - install a pressure relief door

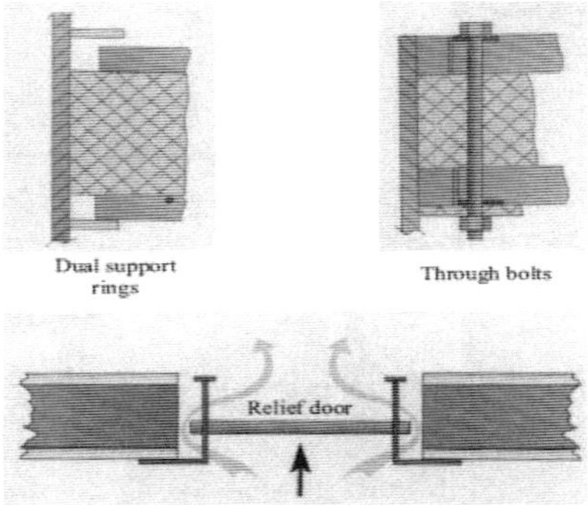

AMACS
PROCESS TOWER INTERNALS

When throughput exceeds original capacity

+ There are various solutions to increase capacity in an existing vessel.

  - Patented High capacity MaxCap® mesh mist eliminator 0% - 100%

AMACS
PROCESS TOWER INTERNALS

When throughput exceeds original capacity cont.

-Patented MISTFIX™
cylindrical mist eliminator 0% - 25%

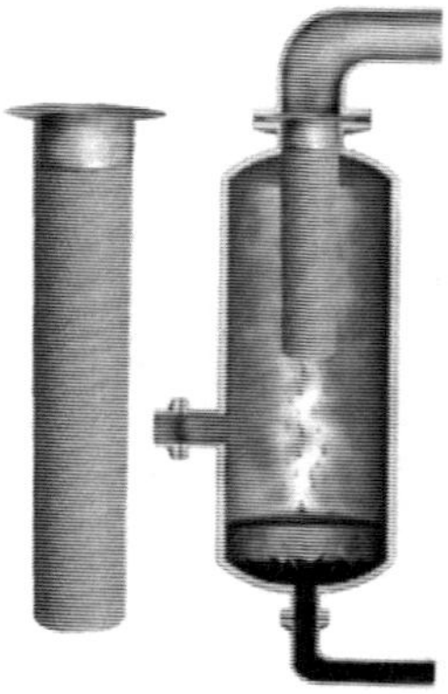

When throughput exceeds original capacity cont.

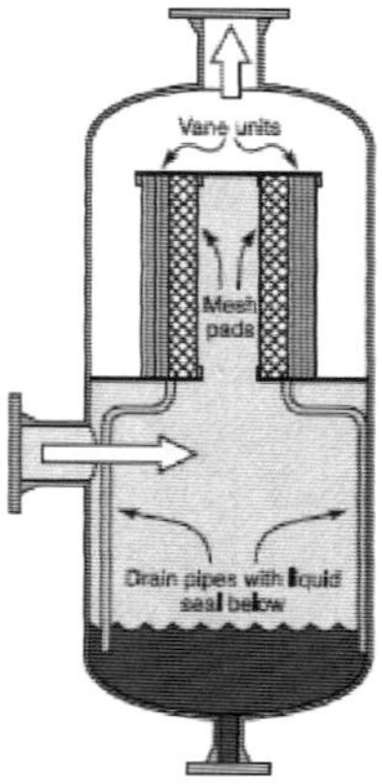

- High capacity Patented MultiPocket® vane. 0%-50%

AMACS
PROCESS TOWER INTERNALS

When throughput exceeds original capacity cont.

+ MistFix™ is installed through the gas outlet, all other M.E. equipment comes in sections, installed through the manway.
+ With increased capacity comes increased bad gas distribution.
  - Use an inlet diffuser.
+ To avoid welding new support rings and housing for the high capacity new equipment, use double expansion rings, a compression fit.

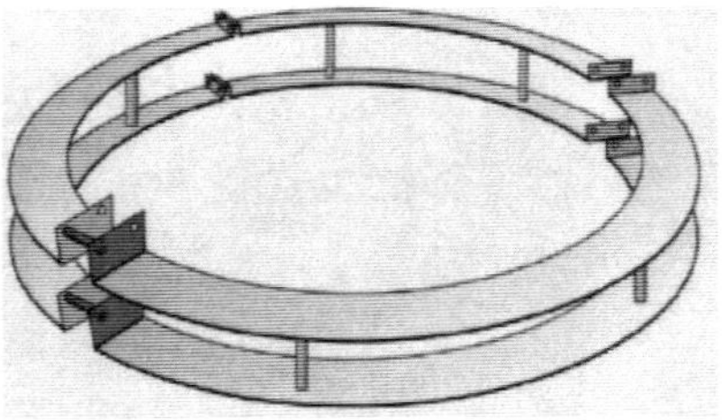

AMACS
PROCESS TOWER INTERNALS

Support Rings

AMACS
PROCESS TOWER INTERNALS

Support Rings

AMACS
PROCESS TOWER INTERNALS

When the capacity range is too large for a suction drum

+ Typical M.E. equipment will operate between

  30% – 110% capacity.

+ When a suction drum often operates below 30% liquid carryover may occur.

AMACS
PROCESS TOWER INTERNALS

When the capacity range is too large for a suction drum Cont.

To facilitate a capacity range of 10% - 110% consider a 2 phase M.E. in the same drum.

- one that operates 30% - 1
- second that operates between 10% - 30%

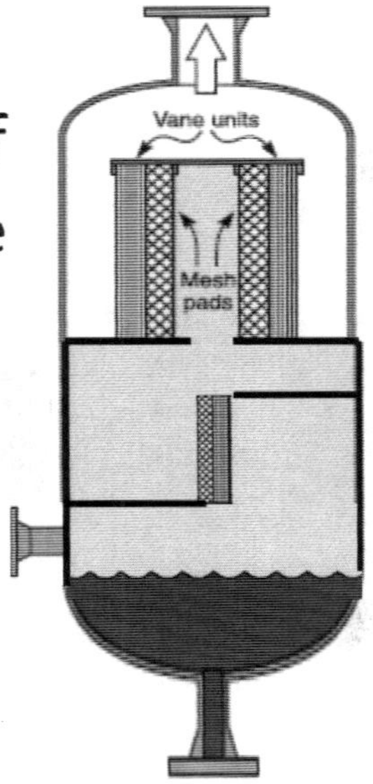

(45i) Renewable Feedstock for Steam Crackers: Catalytic Upgrading of Crude Tall Oil (CTO) Into Bio-Naphtha

Jinto Anthonykutty, Juha Linnekoski, Antero Laitinen and Ali Harlin, VTT Technical Research Centre of Finland, Espoo, Finland

Renewable feedstocks from bio resources are better alternatives to petroleum derived naphtha for the production of olefins and aromatics by steam cracking or reforming.[1, 2] Mostly, these renewable feedstocks (bio-naphtha) rich in n-alkanes are produced by catalytic upgrading methods via hydrotreating/hydroprocessing; in which oxygenates are removed from bio-oil by various deoxygenation processes. At present, hydrotreated vegetable (HVO) oil is produced in many countries on industrial scale using existing petroleum refinery methods[3]. However, employing higher priced vegetable oils could be a major concern in these processes and the main challenge is therefore finding an inexpensive and non-food chain affecting feedstock. Tall oil, obtained as a by-product from Kraft pulping process, is abundant and cheap;and hence, meets the criteria of a low cost starting material for upgrading[4]. Moreover, the utilization of tall oil in conventional oil refineries could serve as an eco-friendly method to convert bio-wastes into bio-naphtha[1].

In the present research approach, catalytic upgrading (hydrotreating) of crude tall oil (CTO) into bio-naphtha suitable for a steam cracker is discussed. CTO, which contains mainly fatty acids (42.9 wt. %) and resin acids (25.7 wt. %) was obtained from Stora-Enso Kaukopää (Finland) mill and used as received. A commercial NiMo catalyst was employed and the experiments were conducted in a continuous, down flow, fixed bed reactor at different process conditions (WHSV 1-3h^{-1}, Pressure 5 MPa and Temperature 350 - 450 ^{o}C). After the stabilization period (6 h.), the liquid sample was collected and separated into organic and aqueous phases.The product distribution in organic phase was analyzed with a Shimadzu GC/MS equipped with a Zebron MS 30m × 0.25 mm × 0.25μm column.

This current approach reports that CTO produces mainly paraffins, cycloalkanes and aromatics during upgrading process. Paraffinic hydrocarbons mainly comprise n-octadecane (20-25 wt. %) and n-heptadecane (15-20 wt. %), are obtained in maximum yield (55- 60 wt. %) in low temperature reactions. The yield of paraffinic hydrocarbons increased as a function of increasing space time in low temperature reactions. 18-Norabietane and other mono-, di- and tri- cycloalkanes are obtained from resin acids in CTO, and these cyclic structures are converted to aromatics such as naphthalene and phenanthrene in high temperature reactions (>400 ^{o}C). Stigmastane, apparently produced from sterols in CTO are also present as a major three ring component in the product stream. Additionally, gaseous products obtained and the yield increased as a function of temperature. The trend observed with the distribution of gaseous products reveal that decarboxylation and cracking reactions are favored at higher temperatures, while hydrogenation and hydrodeoxygenation (HDO) are suppressed.

The elemental composition of CTO shows that it contains relatively high sulfur content (1800 mg/kg), could be useful to keep the catalyst's activity with time on stream (TOS). However, the considerable amount (87 mg/kg) of alkali metals (Na, K, Ca and P) in CTO might be critical for the catalyst activity with TOS, will be discussed in this contribution.

The product composition obtained with CTO is assessed in comparison with compositions and characteristics of refinery naphtha feeds, and will be discussed in detail. Furthermore, this contribution will discuss the achieved deoxygenation rate and success criteria for the product composition of upgraded CTO.

References

1. Gornay, J.; Coniglio, L.; Billaud, F.; Wild, G. Steam Cracking and Steam Reforming of Waste Cooking Oil in a Tubular Stainless Steel Reactor with Wall Effects. ***energy&fuels***, **2009**, 23(11), 5663-5676.
2. Pyl, S.P.; Schietekat, C.M.; Reyniers, M.F.; Abhari, R.; Marin, G.B.; Van Geem, K.M. Biomass to olefins: Cracking of renewable naphtha. ***Chemical Engineering Journal***, **2011**, 176-177, 178-187.
3. Sunde, K.; Brekke, A.; Solberg, B. Environmental Impacts and Costs of Hydrotreated Vegetable Oils, Transesterified Lipids and Woody BTL- A Review. ***Energies***, **2011**, 4(6), 845-877.
4. Coll, R.; Udas, S.; Jacoby, W.A. Conversion of the Rosin Acid Fraction of Crude Tall Oil into Fuels and Chemicals. ***energy&fuels***, **2001**, 15, 1166-1172.

AIChE Paper Number 45j
High Efficiency Ethylene Cracking Furnaces

Rajaram Ramesh

Product Development Manager - Ethylene
Technip
Netherlands

Marco W. M. van Goethem, PhD

Product Development Manager – Modeling and Testing
Technip
Netherlands

25th Ethylene Producers Conference
AIChE 2013 Spring National Meeting
San Antonio, Texas, April 29 – May 3, 2013

Abstract

Environmental and economic pressures generate a driving force to produce products more efficiently. A direct effect of more efficient processes is that the emission of green house gases, particularly carbon dioxide decreases. In households nowadays "high efficiency furnaces" are available with efficiencies in the range of 94-97%. This paper describes the outcome of a feasibility study on applying the high efficiency furnace concept to steam cracking furnaces. The high efficiency is achieved by enhanced heat recovery from the flue gases before they are released to the atmosphere at ca. 65°C.

Technip studied the application of this concept for cracking furnaces based on gas as well as liquid feedstocks. Natural gas was used as basis for the energy price. Since energy prices have a strong impact on project economics, the evaluations were done for an EU average natural gas price and a US natural gas price.

Two alternative heat sinks have been considered: preheating of combustion air and preheating of cold hydrocarbon feed. Both are viable for liquid cracking furnaces. In the case of gas cracking furnaces feed preheating results in a pinch and this makes it not viable.

Capital costs are based on Technip's cost database and quotations from air preheater vendors. Special attention has been paid to the material selection of the affected convection section banks.

This study shows that high efficiency ethylene cracking furnaces are economically viable. For Europe the study indicates an attractive payback period. In the US the payback period is longer due to the lower energy prices. If energy prices increase in the future, the concept's economic viability will become better. Inclusion of CO2 credits will make the payback period more attractive.

Furnace Efficiency

In this paper furnace efficiency has been calculated as net thermal efficiency as interpreted by API560[1]. Furnace efficiency is determined by drawing an envelope around the furnace and including in it the radiant section, convection section and any heat exchanger equipment that recover heat from the flue gas (air preheater, economizer, steam superheater, etc.). The efficiency is the ratio of the useful heat absorbed to the fired heat of the fuel and any other external heat inputs. Ambient temperature is a reference temperature for efficiency calculations and any heat inflow or outflow above this reference is taken into consideration as a heat input or heat loss.
Furnace efficiency can be determined as follows:

$$\frac{heat\ absorption\ in\ radiant\ section\ +\ convection\ section}{(firing\ rate\ \times\ fuel\ heating\ value) + sensible\ heat\ of\ fuel\ and\ air\ above\ ambient\ temp. + other\ external\ heat\ inputs}$$

Alternately, Furnace efficiency $= 100\%$ $-$ Total heat loss, where:

$$\text{Total heat loss} = \frac{stack\ loss\ +\ radiation\ heat\ loss}{(firing\ rate\ \times\ fuel\ heating\ value) + sensible\ heat\ of\ fuel\ and\ air\ above\ ambient\ temp. + any\ external\ heat\ inputs}$$

In a furnace where fuel and combustion air enter the furnace at ambient temperature and there are no other external heat inputs, the first expression can be simplified to the following:

$$Furnace\ efficiency\ = \frac{heat\ absorption\ in\ radiant\ section\ +\ convection\ section}{fuel\ firing\ rate\ \times\ fuel\ heating\ value}$$

Please note that net thermal efficiency, as defined by API 560, is based on the lower heating value of the fuel.

Study objectives

Technip undertook this study with the goal to reduce fuel firing in cracking furnaces by at least two percent. Although this figure does not look very ambitious, it must be remembered that since efficiencies in current designs are already in the mid nineties (upper part of the technology S-curve), further improvements will not be easy and can only be incremental.

Since stack loss is the main source of inefficiency (radiation heat loss is comparatively smaller and more difficult / expensive to reduce) it was decided to make stack loss reduction as the main focus of the study.

By a relatively straight forward calculation it can be shown that the two percent efficiency rise target would require the flue gas to be cooled to approximately sixty or seventy degrees Celsius.

The main challenge in reducing stack temperature to this level is finding an appropriate heat sink for the recovered heat. In a cracking furnace there are two suitable heat sinks, combustion air and hydrocarbon feed, and both were studied. The following sections discuss the results for liquid furnaces and gas furnaces.

Basis of Study

We have done the study on a naphtha furnace and an ethane furnace. We have chosen these two feedstocks since a large number of furnaces operate on these two feeds and we can draw conclusions that can be applied approximately to other liquid and gas feeds. The study has been performed on the following bases:

- The reference liquid and gas furnace designs used in this study are based on actual furnaces from Technip's past projects. The furnaces were simulated using Technip's proprietary tool SPYRO which is used in the design and simulation of cracking furnaces.
- The budgetary capital cost need to achieve the target furnace efficiency was calculated as an incremental cost. It was estimated from vendor quotes and Technip's internal cost database.
- The cost estimates are for a new project, not a revamp, and are for a West European location.
- Operating costs are based on cost of base utilities and derived utilities. Base utilities are natural gas and demineralized water. These are taken from market data.
- Since natural gas prices are very different in different parts of the world, the operating costs and economic analysis have been done for European and United States gas prices.
- EU natural gas price was fixed at €25.1 /MWh of higher heating value (HHV) which is equivalent to €28.0 /MWh of lower heating value (LHV). The prices are based on EUROSTAT data for 1st half of 2010 and were obtained from the publications of the Department of Energy & Climate Change in the UK. They correspond to the EU-15 average for Large Industrial Consumers (i.e. 2.13 pence/kWh HHV).
- US Natural gas prices were obtained from the US Department of Energy. The average value was 6 \$/MM BTU (HHV) or 17 €/MWh (LHV) for 1st half of 2010. The average HHV used is 1029 BTU per cubic foot or 1.029 MM BTU per thousand cubic feet.
- Cost of demineralized water was assumed at € 2.00 /ton.
- The prices of other utilities (boiler feed water, steam and electricity) have been calculated from those of the two base utilities. Electricity was priced assuming 35% overall efficiency of power generation.
- The economic calculations do not include any CO2 credit for the fuel saved.
- Assumed exchange rate: 1 US dollar = 0.75 euros
- The plant is assumed to operate 8400 hours per year

The resultant utility prices used in the study are summarized below.

UTILITY COSTS

		EU basis	US basis	
FUEL as NATURAL GAS	Base	**28.0**	**17.0**	€ / MWh
DEMINERALISED WATER	Base	2.0	1.5	€ / ton
VHP STEAM (net value)	Derived	23.7	14.4	€ / ton
MP STEAM (net value)	Derived	13.9	8.5	€ / ton
LP STEAM	Derived	9.7	5.9	€ / ton
ELECTRICAL POWER	Derived	80.1	48.6	€ / MWh

Study Results

Naphtha Furnace using Air Preheat

Combustion air can be preheated in a plate type air preheater (APH). These are cross flow type exchangers and have low pressure drops. Combustion air is at a lower temperature than any of the process or utilities streams entering the convection section; therefore, from the standpoint of efficiency, APHs are ideally suited to be the last heat recovery section before the stack.

Fig 1 shows the Base Case Naphtha furnace while Fig 3 shows the same furnace with an APH. Please also refer to Table 1 for a performance comparison of the APH design with the Base Case naphtha furnace. The following points are of interest:

1) The overall efficiency of the furnace will improve from 94.4 to 97.0% for the APH option. This is mainly due to the lower stack temperature but also partly due to the smaller excess air, which is explained below.

2) This option will require a forced draft combustion air fan to overcome the pressure drop in the APH. As a result the furnace operation will change from induced draft mode to balanced draft. The availability of air in forced draft will enable more accurate control of the excess air and it will be possible to operate the furnace with lower excess air (5% instead of 8%). This leads to a lower flue gas flow, a lower stack loss and therefore better efficiency.

3) There will be a drop of 4.8% in fuel firing which is greater than the efficiency increase. This is because of the lower absorbed heat. The furnace is designed and controlled in such a way as to supply all the required process heat and to absorb the surplus heat into utilities to the extent possible. When the furnace operates more efficiently and the heat demanded by the process media remains unchanged, as in the APH case, there will be less surplus heat available for VHP steam. As a result the VHP steam generation will go down from 79.7 to 76.0 t/h.

4) The heat transfer surface in some parts of the APH will be below the water dew point of the flue gas (around 60°C). This can lead to water condensation and create two possible issues: 1) accumulation of water in the APH and 2) corrosion due to the presence of liquid water, oxygen and carbon dioxide.

   Two measures have been included in the design to counter these problems:
   - Placement of unit such that flue gas flows downward to aid easy draining
   - Use of corten steel to resist corrosion

5) We have evaluated the impact of the air preheat option on NOx emissions using API 535[2)]. The rise in air temperature and the drop in excess air will have opposing effects on the NOx production. The net effect is that the NOx concentration will be close to the Base Case.

   If a DeNOx unit is present, as is often the case in the US, the APH option may help to slightly extend the life of the DeNOx catalyst. This is because the total NOx load will be smaller than the Base Case due to the smaller flue gas flow.

Table 2 summarizes the results of the economic analysis of this option. We have used payback period to compare different options. Payback period is the ratio of the incremental capital cost above the Base Case CAPEX divided by the savings in yearly operating cost.

1) The capital cost (CAPEX) of the APH option includes the cost of the APH, the combustion air fan, extra ducting and structural steel. The operating cost (OPEX) savings is the savings in furnace fuel but this must be offset against the value of the lost VHP steam production (which notionally must be made up in a utility boiler elsewhere) and the additional electricity for the two fans.

2) It can be seen that the payback period is three years based on European gas prices and five years based on American gas prices.

Naphtha Furnace using Feed preheat

Naphtha is normally stored at ambient temperature and is preheated to 60°C using quench water before it is routed to the furnaces. Most cracking furnaces are equipped with a Feed Preheater (FPH) bank in the convection section. If flue gas is utilized to preheat naphtha from ambient conditions the stack temperature can be lowered to the desired range of sixty to seventy degrees Celsius. Till now designers and operators have not taken full advantage of this mainly because of the higher CAPEX and fear of dew point corrosion.

One of the consequences of this feature will be that the quench water duty previously absorbed by the naphtha will now shift to the air coolers in the quench water circuit. The additional electrical power that the air cooler fans will require will be about 10 KW per furnace.

Please refer to Table 3 for an overview of the performance of a naphtha furnace with and without an extended feed preheater. Please note that the Base Case naphtha furnace is the same as the one considered in the APH option.

1) The overall efficiency of the furnace can be improved from 94.4 to 96.0% for this option.

2) Unlike the APH option a combustion air fan will **not** be required. The furnace will continue to operate in induced draft mode.

3) The fuel firing and the VHP steam generation will drop marginally.

4) The upper tube rows of the FPH will be below the water dew point of the flue gas and some condensation will occur. As a precaution we have considered that the top four rows will be of stainless steel.

5) The flue gas NOx concentration will be unaffected by this option. However the total hourly NOx emissions and load on the DeNOx unit, if any, will decrease to the same extent as the fuel firing, which is around 0.6%.

Table 4 summarizes the results of the economic analysis of this option.

1) The capital cost (CAPEX) of this option covers the incremental cost of the extended FPH, extra ducting and structural steel. The OPEX savings is the reduction in furnace fuel after adjusting for the reduction in VHP steam production and the additional electricity for the ID fan and the Quench Water Air Cooler fans.

2) In the Base Case the naphtha feed piping between the external feed preheater and the convection section is insulated for heat conservation. In the FPH option naphtha will be fed at ambient temperature to the furnace and so these lines do not have to be insulated. This will yield some capital cost saving but this has not been taken into account in our analysis.

3) The payback period excluding CO2 credit will be three and a half years with European natural gas and five and a half years with US natural gas.

Ethane Furnace using Air Preheat

Fig 2 shows the Base Case Ethane furnace that was used in the study. The APH arrangement on an ethane furnace will be quite similar to the one on the naphtha furnace. Table 5 shows the impact that the APH would have on the furnace performance. The main points are as follows:

1) The overall efficiency of the furnace will improve from 91.7 to 96.5% for the APH option. This is mainly due to the lower stack temperature but also partly due to the smaller excess air.

2) In this furnace also the APH will require a forced draft combustion air fan. This will allow the excess air to be reduced from 8% to 5% which will result in lower stack losses and a better efficiency.

3) The fuel firing will go down by 11.0%, from 8.2 to 7.3 t/h. The absorbed heat will drop from 116.8 to 110.6 MW. As a result of these the VHP steam generation will go down from 97.0 to 88.8 t/h.

4) A portion of the APH will operate below the flue gas water dew point. Consequently the unit must be designed to prevent water accumulation and corrosion.

5) The impact on NOx was evaluated here also. The rise in air temperature will have a greater impact on NOx than the fall in excess air. We estimate that as a result the flue gas NOx concentration will increase slightly (around two percent). This can be a cause for concern if the furnace does not have a DeNOx unit and is operating close to the limits imposed by authorities. However since the total firing will go down, the hourly NOx production will drop. Plants can consider discussing with authorities to obtain a higher allowable NOx limit,

   If the furnace has a DeNOx unit, the chemical load on the unit will reduce, which should help to obtain a longer catalyst life.

The operating cost savings and payback for this option have been summarized in Table 6.

1) The capital cost includes the cost of the APH, the combustion air fan, extra ducting and structural steel. The operating cost savings is the savings in fuel firing with an adjustment for the value of the reduced VHP steam generation and the extra electricity for the ID fan and FD fan.

2) The payback period with natural gas priced at European rates will be one and a half years. The corresponding figure for US gas prices will be three years.

Ethane Furnace using Feed Preheat

Using feed at ambient temperature does not significantly reduce the stack temperature in a gas furnace. This is because the feed preheat curve has a steeper slope than the flue gas cooling curve, causing a pinch on the hot side of the Feed Preheater. Fig 4 shows the heating and cooling curves of ethane, boiler feed water and flue gas in the convection section. The hatched shape on the plot shows the maximum additional heat duty that can be recovered into feed, over and above the Base Case. The plot clearly shows that when the ethane feed temperature is lowered from the base value of 55°C to 30°C, the stack temperature will drop from 170°C to about 150°C. The flue gas cannot be cooled any more than this modest amount.

For this reason the feed preheat option was not investigated any further.

Conclusions and Summary

The key results from the different study options are listed together below.

Cracking furnace feed	Naphtha		Naphtha		Ethane	
Cold sink	Combustion air		Hydrocarbon feed		Combustion air	
	EU nat. gas	US nat. gas	EU nat. gas	US nat. gas	EU nat. gas	US nat. gas
OPEX savings	€ 0.60 mill	€ 0.36 mill	€ 0.17 mill	€ 0.10 mill	€ 1.27 mill	€ 0.77 mill
Payback time	3 years	5 years	3.5 years	5.5 years	1.5 years	3 years
Salient features of each option	Small number of additional equipment		No additional equipment		Small number of additional equipment	
	Higher CAPEX		Lower CAPEX		Higher CAPEX	
	More OPEX savings		Less OPEX savings		Best OPEX savings	
	Slight effect on NOx concentration		No effect on NOx		Around 2% higher NOx concentration	

All three scenarios have attractive payback periods and will be worth considering in high energy cost countries such as Europe and Asia. In US and other energy rich nations, the economic incentive will be much lower. In the US, where ethane cracking is common and increasing even more, improvement of furnace efficiency with air preheat can be worth considering as the payback time is less than 3 years.

If the options are compared against each other, in our opinion the Air preheating option has more potential, irrespective of feedstock.

The Feed Preheating option in liquid furnaces is cheaper but it also gives proportionally modest savings. Moreover, as earlier mentioned, the feed preheat duty is being shifted from one waste heat source (quench water) to another (flue gas). This option will not lead to a net improvement or cash savings in the plant as a whole unless there is a shortage of quench water and there are other, better uses for the saved quench water duty

Acknowledgements

The authors wish to express their thanks to their colleagues Mr Ashok Laghate, Ms Susan Cooper, Mr Sachin Singare and others in Technip Benelux who contributed to this study.

References

1) API 560: Fired Heaters for General Refinery Service, 4th edition

2) API 535: Burners for Fired Heaters in General Refinery Service, 2nd edition

Table 1
Naphtha Furnace with Combustion Air Preheating – Performance Summary

		Base Case	Air Preheat Case
Naphtha feed	t/h	60.8	60.8
Ambient conditions	-	15°C, 85% R.H.	15°C, 85% R.H.
Air temp. to burners	°C	15	77
Stack temperature	°C	108	61
Excess air	%	8	5
Absorbed heat	MW	113.2	110.4
Overall efficiency	%	94.4	97.0
Fuel flow	t/h	8.4	8.0
Fired heat	MW	119.9	113.9
VHP steam production	t/h	79.66	76.01
NOx concentration		Base	Close to Base

Table 2
Economics of Naphtha Furnace with Combustion Air Preheating

		Air Preheat Case (EU location)	Air Preheat Case (US location)
Savings in fuel firing	million €/yr	1.42	0.86
Cost of lost VHP steam capacity	million €/yr	0.73	0.44
Cost of additional electricity	million €/yr	0.10	0.06
Net OPEX saving	million €/yr	0.60	0.36
Payback	Years	3	5

Table 3
Naphtha Furnace with Feed Preheating – Performance Summary

		Base Case	Feed Preheat Case
Naphtha feed	t/h	60.8	60.8
Feed temperature	-	60°C	30°C
Combustion air temp.	°C	15	15
Stack temperature	°C	108	71
Excess air	%	8	8
Absorbed heat	MW	113.2	114.3
Overall efficiency	%	94.4	96.0
Fuel flow	t/h	8.4	8.35
Fired heat	MW	119.9	119.1
VHP steam production	t/h	79.66	79.63
NOx concentration		Base	Base

Table 4
Economics of Naphtha Furnace with Feed Preheating

		Feed Preheat Case (EU location)	Feed Preheat Case (US location)
Savings in fuel firing	million €/yr	0.19	0.12
Cost of lost VHP steam capacity	million €/yr	< 0.01	< 0.01
Cost of additional electricity	million €/yr	0.01	< 0.01
Net OPEX saving	million €/yr	0.17	0.10
Payback	Years	3.5	5.5

Table 5
Ethane Furnace with Air Preheating – Performance Summary

		Base Case	Air Preheat Case
Ethane feed	t/h	51.9	51.9
Ambient conditions	-	15°C, 85% R.H.	15°C, 85% R.H.
Air temp. to burners	°C	15	129
Stack temperature	°C	169	64
Excess air	%	10	5
Absorbed heat	MW	116.8	110.6
Overall efficiency	%	91.7	96.5
Fuel flow	t/h	8.2	7.3
Fired heat	MW	127.3	114.5
VHP steam production	t/h	97.0	88.75
NOx concentration		Base	Base + 2%

Table 6
Economics of Ethane Furnace with Air Preheating

		Air Preheat Case (EU location)	Air Preheat Case (US location)
Savings in fuel firing	million €/yr	3.01	1.83
Cost of lost VHP steam capacity	million €/yr	1,64	1.00
Cost of additional electricity	million €/yr	0.10	0.06
Net OPEX saving	million €/yr	1.27	0.77
Payback	Years	1.5	3

Fig 1
Naphtha cracking furnace

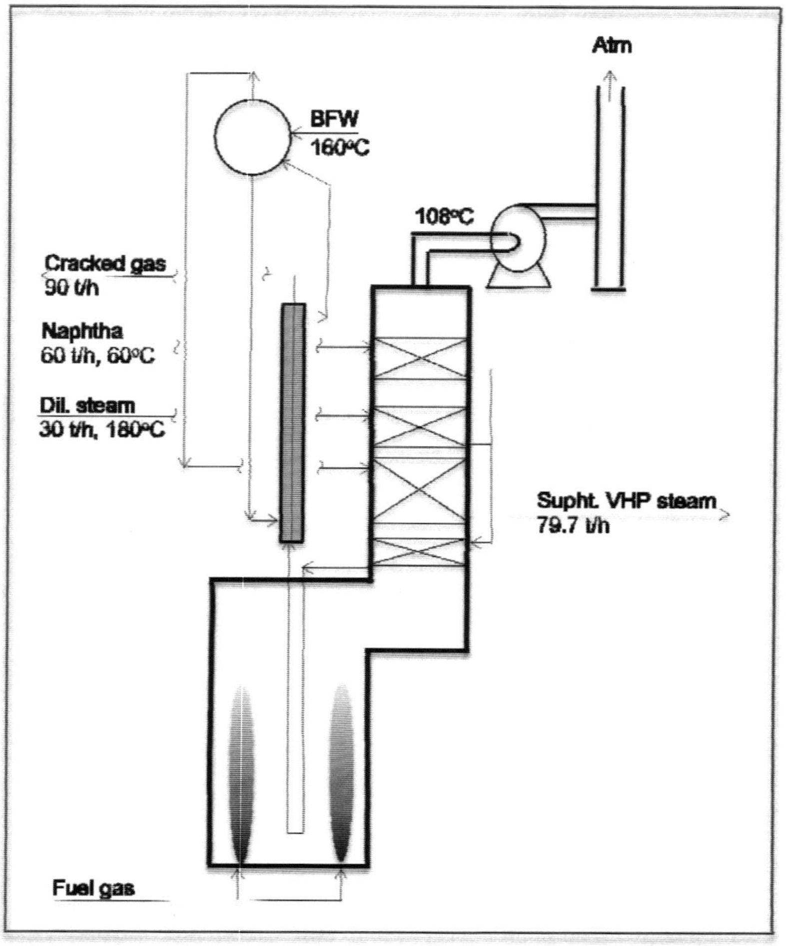

Fig 2
Ethane cracking furnace

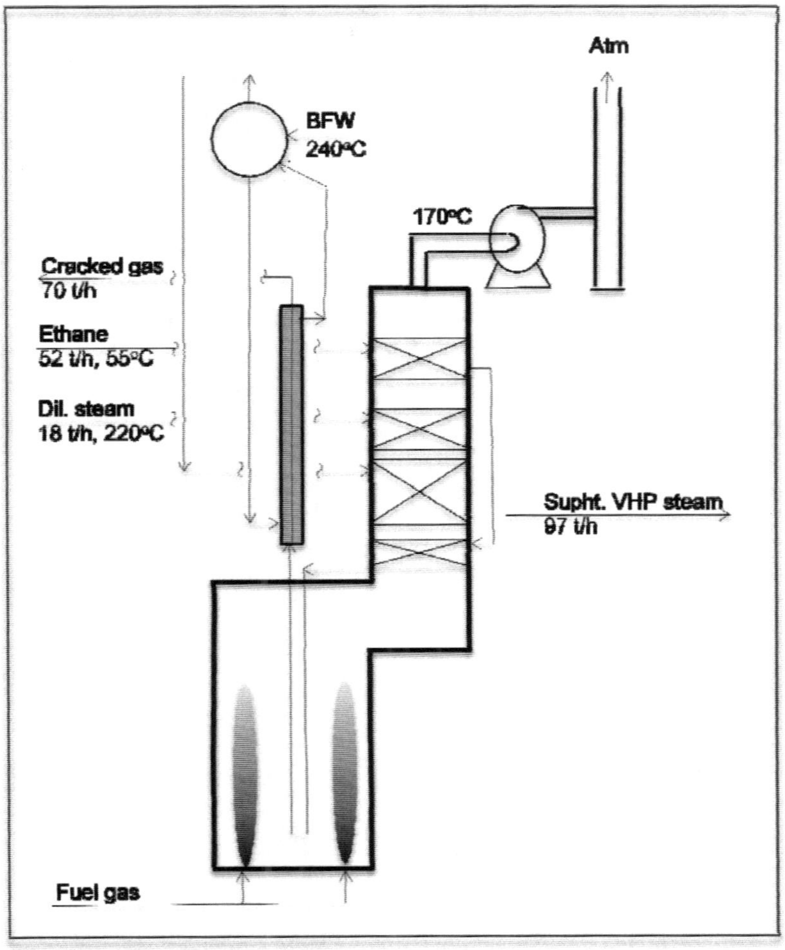

Fig 3

<u>Naphtha cracking furnace – Air Preheater option</u>

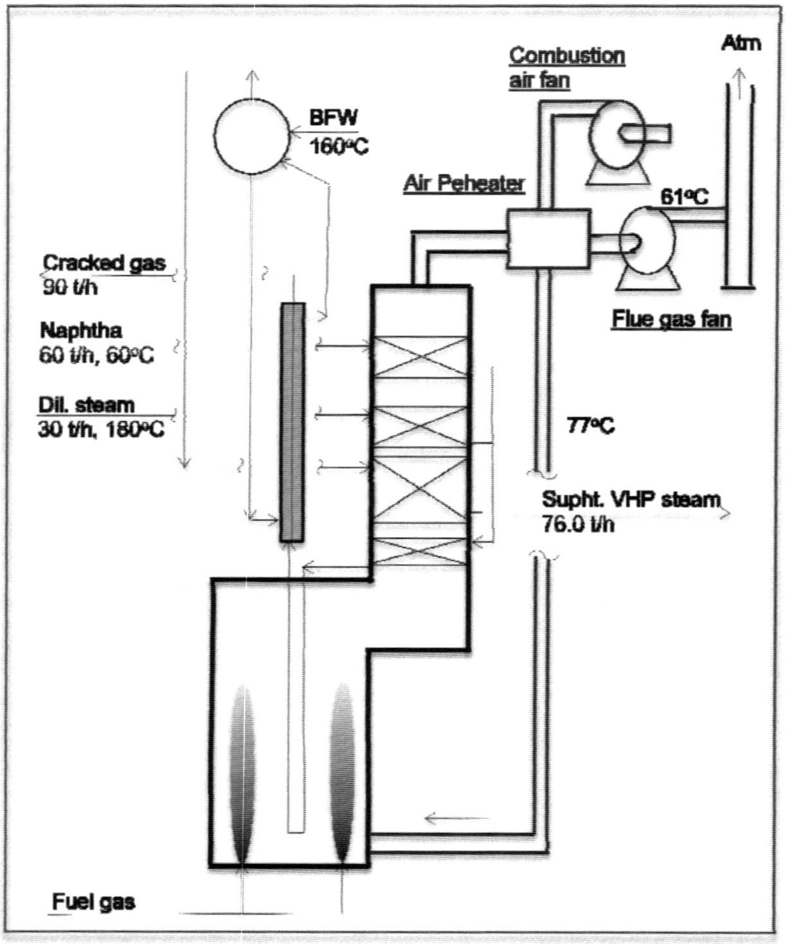

Fig 4

Ethane cracking furnace – Feed Preheater option

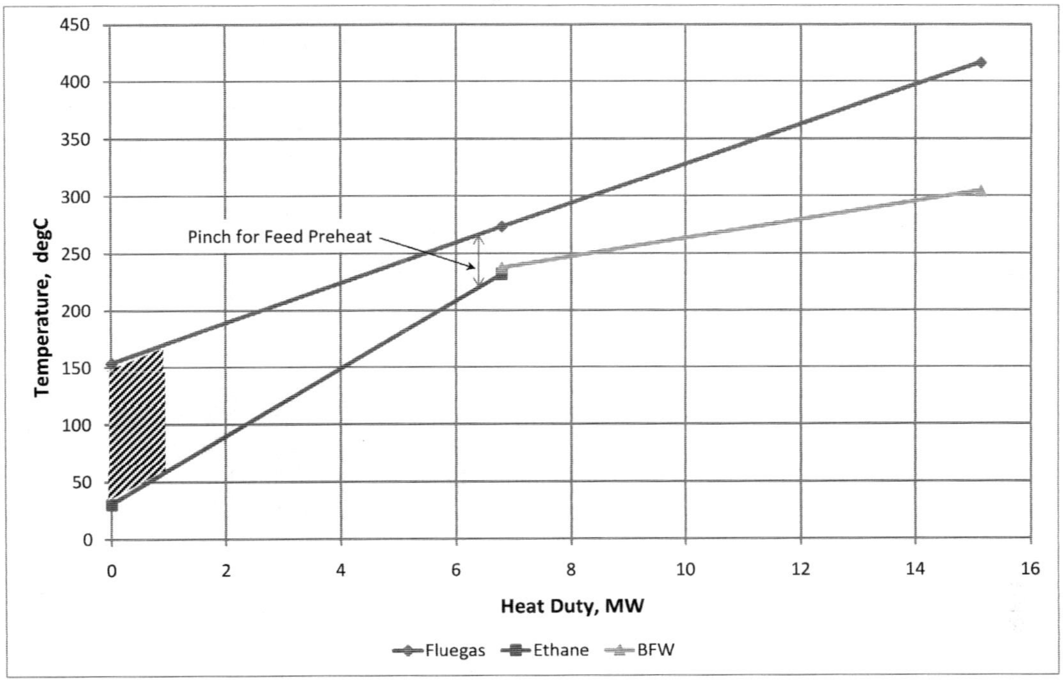

Modernize Your Control Functionality
Easy Steps to Start Process Optimization

13 AIChE
Spring Meeting, San Antonio, TX

Presenters

- Laurie Ben
- James Beall

EMERSON
Process Management

Agenda

- Introduction
- Examples – Best Practices & Lessons Learned
  - Split Range Duty
  - Smith Predictor
  - Gain Scheduling/Adaptive Control
  - Signal Characterization
- Summary

Introduction

- Modernizing Control System – Decision Done Check!
- 'Do I' or 'Don't I' take advantage of new technology
  - Implemented older cumbersome techniques?
  - Complex custom strategies that handle automation challenges?
  - Hard to implement & maintain?
- Standard features that are field proven, practical, and easy to implement
  - Easier to implement & maintain
  - More efficient and effective control strategies
  - Achieve improvements in process performance
  - Lays the foundation for optimization

Agenda

- Introduction
- Examples – Best Practices & Lessons Learned
  - Split Range Duty
    - Control Challenge
    - Legacy Solution
    - Newer Technology Solution
    - Summary
  - Smith Predictor
  - Gain Scheduling/Adaptive Control
  - Signal Characterization
- Summary

Split Range Duty – Control Challenge

- Use steam and hot oil as energy source for distillation column for temperature controller.
- Hot oil is "free energy" and should be maximized
- Control Scenarios
  - Maximize hot oil and supplement with steam
  - Base load steam (if meter bad) and supplement with hot oil
  - Base load hot oil flow (if bad meter) and supplement with steam
- Constraints
  - Minimum and maximum hot oil valve position and flow rate
  - Minimum steam flow rate

Split Range Duty – Legacy Solution

- Old legacy system implemented this feature using a velocity PID with incremental output and a "micro expert rules system"

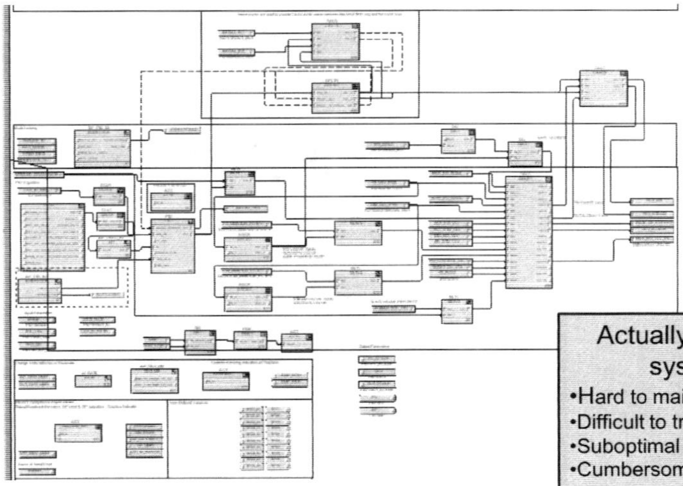

Actually Duplicated in New system, but still…

- Hard to maintain and modify
- Difficult to troubleshoot
- Suboptimal performance
- Cumbersome/Kludge/Ugly

Split Range Duty – New Technology Solution

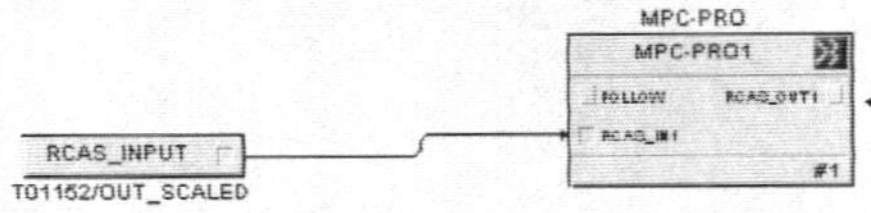

Embedded MPC Function Block provides all control functionality! EASY!!!

Simple calculations to calculate limits

Split Range Duty – New Technology Solution

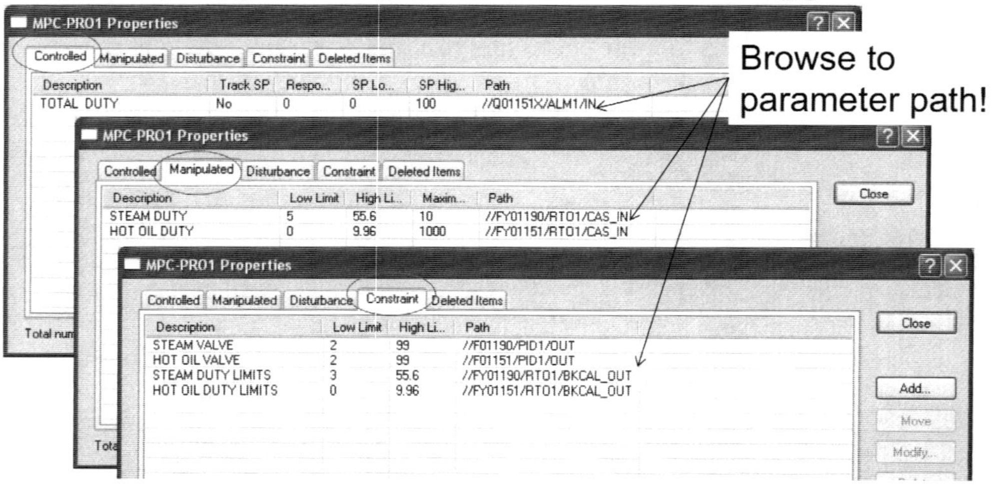

Split Range Duty – New Technology Solution

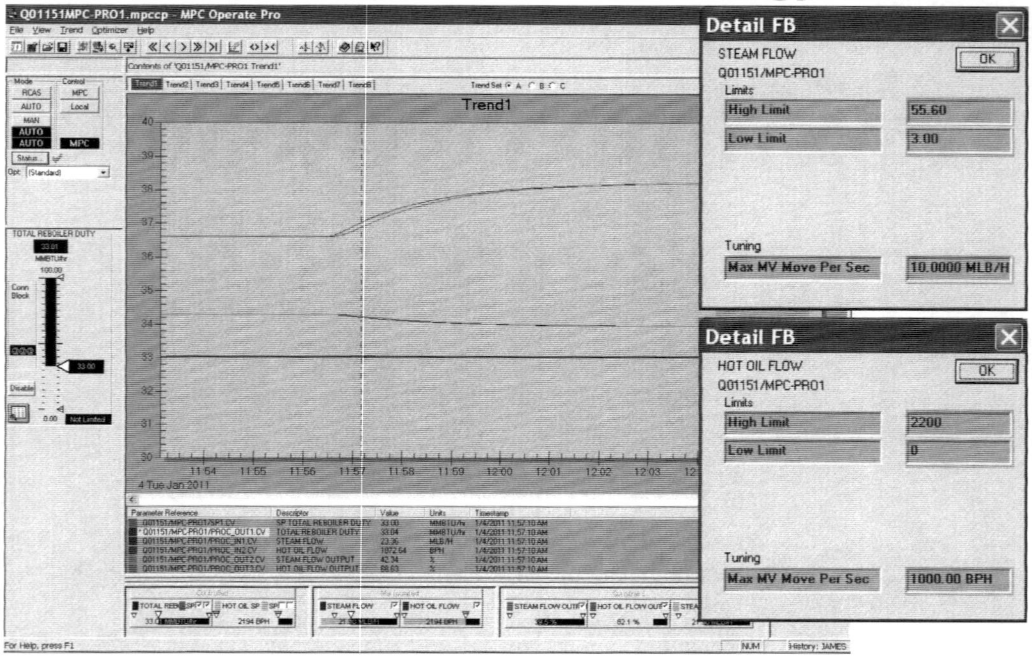

Summary Split Range Duty Control

Embedded MPC (New Technology) vs Legacy Solution

- Better performing solution
  - Bumpless transfers and constraint control
- Easier to Maintain
  - Change limits and settings
  - Simpler implementation
  - Easy to understand, train others, document
- Easily expands to handle future constraints
- Reduce energy costs!

Agenda

- Introduction
- Examples – Best Practices & Lessons Learned
  - Split Range Duty
  - Smith Predictor
    - Control Challenge
    - Legacy Solution
    - Newer Technology Solution
    - Summary
  - Gain Scheduling/Adaptive Control
  - Signal Characterization
- Summary

Smith-Predictor PID – Control Challenge

- Control of processes that are "dead time dominant" (dead time > time constant)
- Typically custom configured
  - some legacy control systems have preconfigured options, e.g. PROVOX
- Algorithm uses a process model
  - (gain, dead time and time constant)
- Challenge: as process dynamics change from model used, performance degrades rapidly

Smith-Predictor PID – Legacy Solution

- Traditional Smith-Predictor

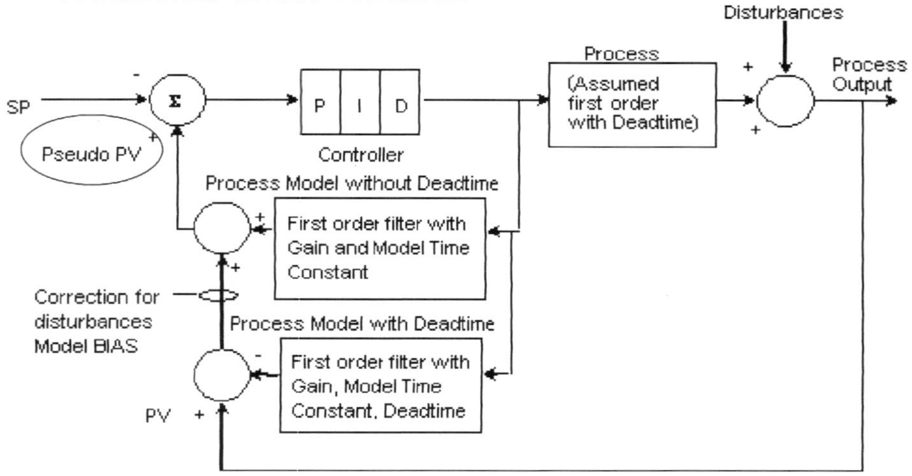

•High maintenance when process dynamics change
•Performance degrades rapidly as process dynamics change from configured model

Smith-Predictor PID – New Technology Solution #1

Easy to implement the Smith-Predictor but…

- High Maintenance when process dynamics change
- Performance degrades rapidly as process dynamics change from configured model

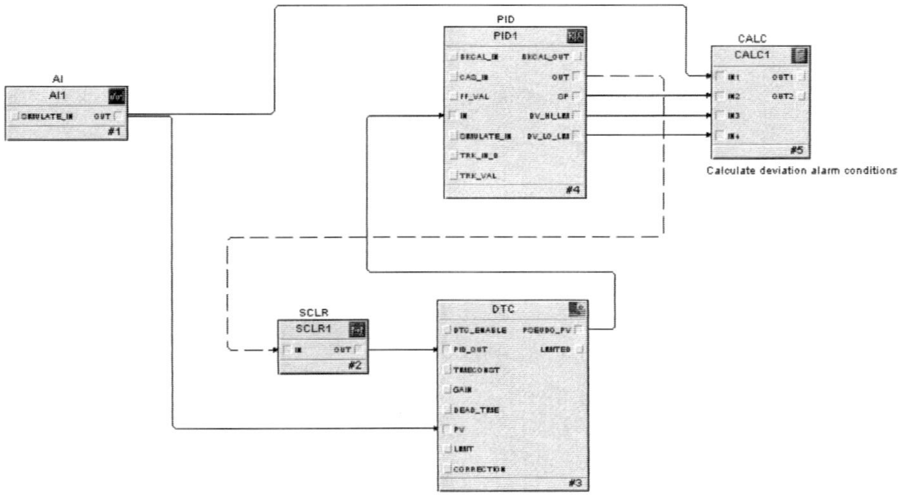

Smith-Predictor PID – New Technology Solution #2

- Embedded MPC Solution

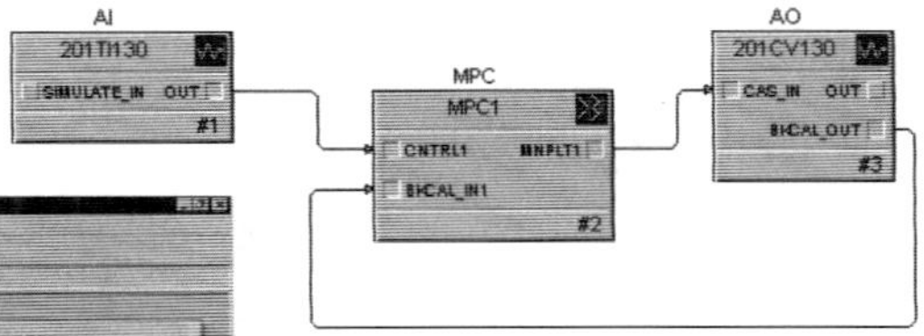

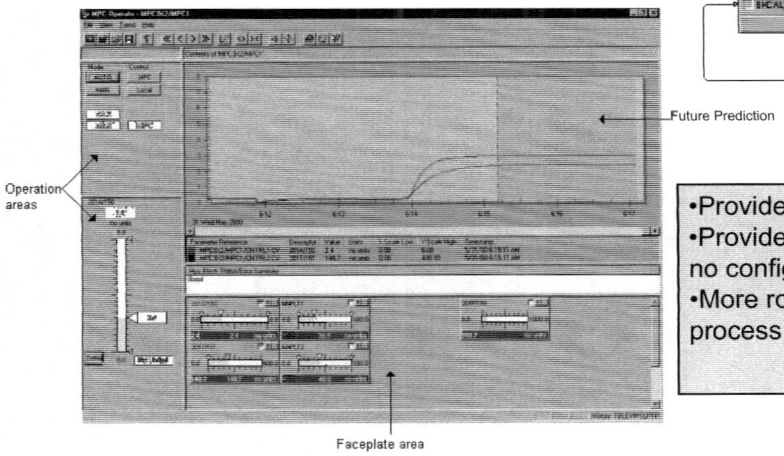

- Provides Future Prediction
- Provides MPC Faceplate – no configuration needed
- More robust under changing process dynamics

Smith-Predictor PID – New Technology Solution #2

- Embedded MPC Solution

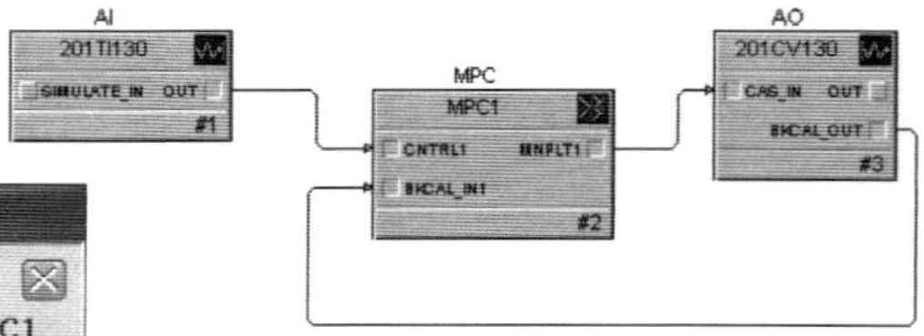

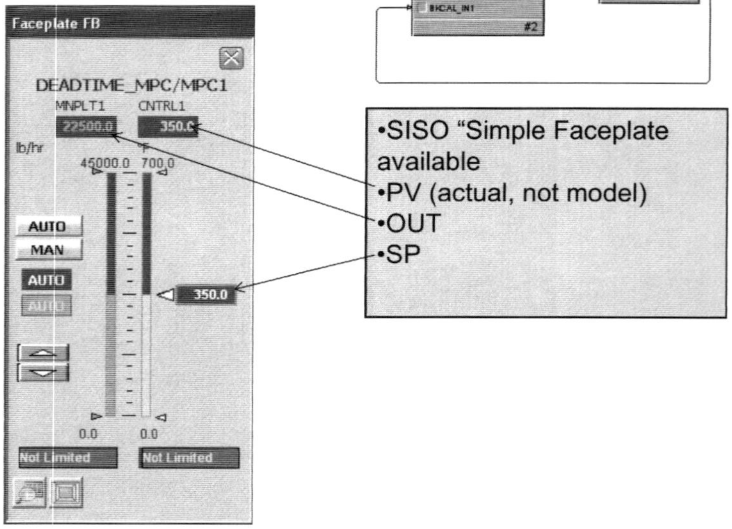

Summary Smith-Predictor PID

New Technology Solutions

- Standard Smith Predictor (#1) and Embedded MPC (#2)
- Both easy to implement and understand!

New Technology Solution #2 (Embedded MPC)

- More robust solution under a changing process model
- Provides visual display of the "future" PV
- Save money through better control performance

Agenda

- Introduction
- Examples – Best Practices & Lessons Learned
  - Split Range Duty
  - Smith Predictor
  - Gain Scheduling/Adaptive Control
    - Control Challenge
    - Legacy Solution
    - New Technology Solution
    - Summary
  - Signal Characterization
- Summary

Gain Scheduling – Control Challenge

- Legacy systems have limited capability to modify tuning based on process conditions
- Others had add-on application software, e.g. PROVOX "Gain Scheduler" system provided 3 sets of Gain, Reset, and Rate applied in 3 separate regions of a "state variable"
  - State variable = PV, OUT or "other"
  - Transition between tuning sets interpolated over "dead band"
- Beware: some suspend loop execution when tuning parameters change, so don't change the tuning every scan

Gain Scheduling – Legacy Solution

- Custom code
- Multiple points, registers, displays
- Restricted (slower) update rate

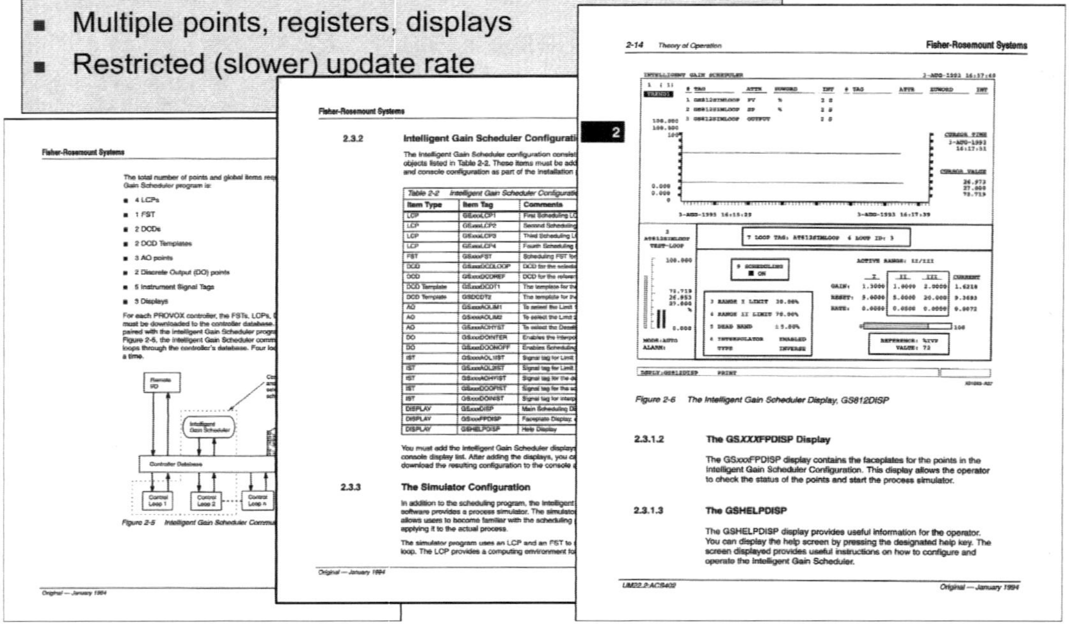

Gain Scheduling – New Technology Solution #1

Standard Modules

- Standard configuration
- Preconfigured faceplate & detailed display
- Easy to implement
- Tuning updates at loop execution

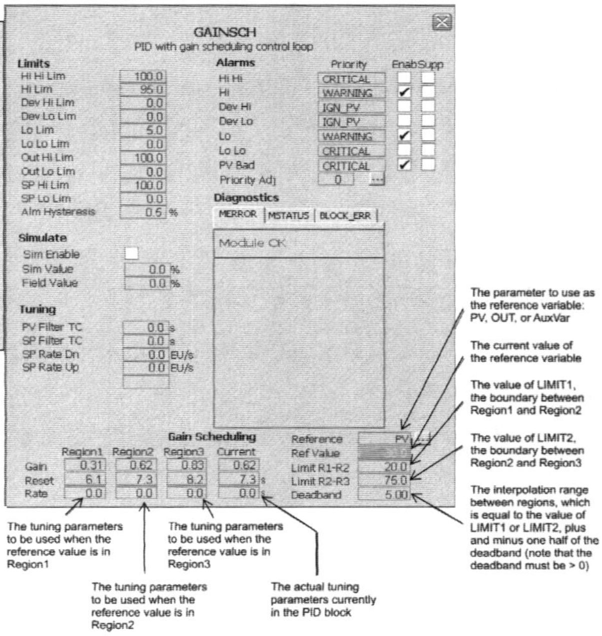

Gain Scheduling – New Technology Solution #2

New Technology "Adaptive Control"

- "Learns" process models from normal operating moves
- Allows up to 5 tuning regions of the "state variable"
- Partial Adaptive Control
  - Use "approved model" only & selected tuning rule in each region
  - Lambda PI or IMC PID; "Closed loop time constant" or "Lambda Factor".
- Optional Full Adaptive Control
  - continuous "Learning" of process model in each region
  - updates tuning while in each region per selected tuning rule
  - set allowable deviation (%) from approved model in each region

New Technology – Adaptive Control

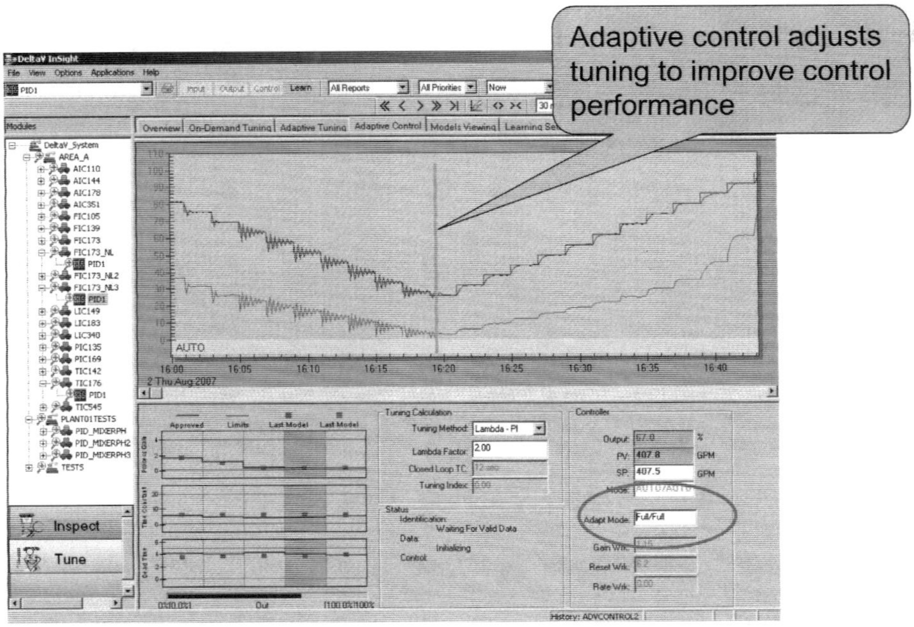

Summary Gain Scheduling

New Technology Solution(s) vs. Legacy Solution

- Gain Scheduling easy to implement and understand!
  - Standard configuration
  - Preconfigured faceplate & detailed display
  - Tuning updates at loop execution
- Adaptive Control provides additional benefit and ease of implementation
  - Learns models during normal operation – NO STEP TESTs!
  - More Regions – Up to 5
  - Automatic tuning updates when process model changes
- Improved control performance saves money!

Summary Gain Scheduling

New Technology Solution(s) vs. Legacy Solution

- Gain Scheduling easy to implement and understand!
  - Standard configuration
  - Preconfigured faceplate & detailed display
  - Tuning updates at loop execution
- Adaptive Control provides additional benefit and ease of implementation
  - Learns models during normal operation – NO STEP TESTs!
  - More Regions – Up to 5
  - Automatic tuning updates when process model changes
- Improved control performance saves money!

Agenda

- Introduction
- Examples – Best Practices & Lessons Learned
  - Split Range Duty
  - Smith Predictor
  - Gain Scheduling/Adaptive Control
  - Signal Characterization
    - Control Challenge
    - Legacy Solution
    - New Technology Solution
    - Summary
- Summary

Signal Characterization - Control Challenge

- Input Characterization (AI)
  - Combustion control – characterize fuel and air
  - Level – characterize pv (e.g. level to volume)
- Output characterization (AO)
  - modify installed flow characteristics of valves, dampers or other final control elements
  - Use to modify loop gain, e.g. output path to slave loop
- Legacy System Challenges
  - Lack of resolution (few points)
  - External black box characterizers may shift, lose documentation
  - Potential operator/maintenance confusion on signal going to final control element

Signal Characterization – Legacy Solution

- Legacy system using external analog characterizer

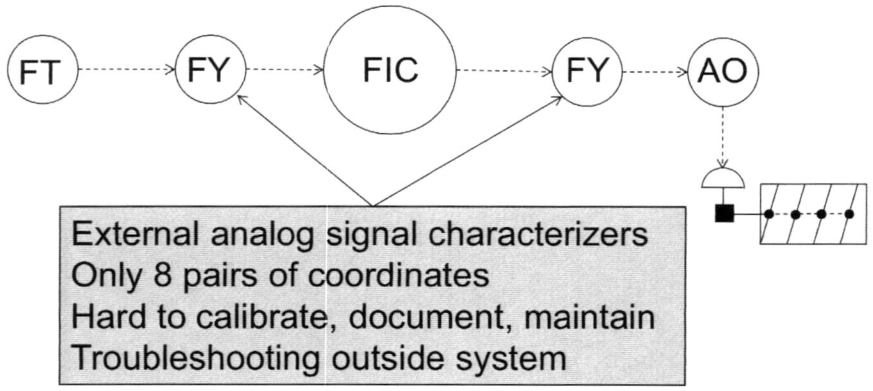

Signal Characterization – New Technology Solution #1

- Standard Module implementation provides 21 point signal characterizer function block (SGCR)

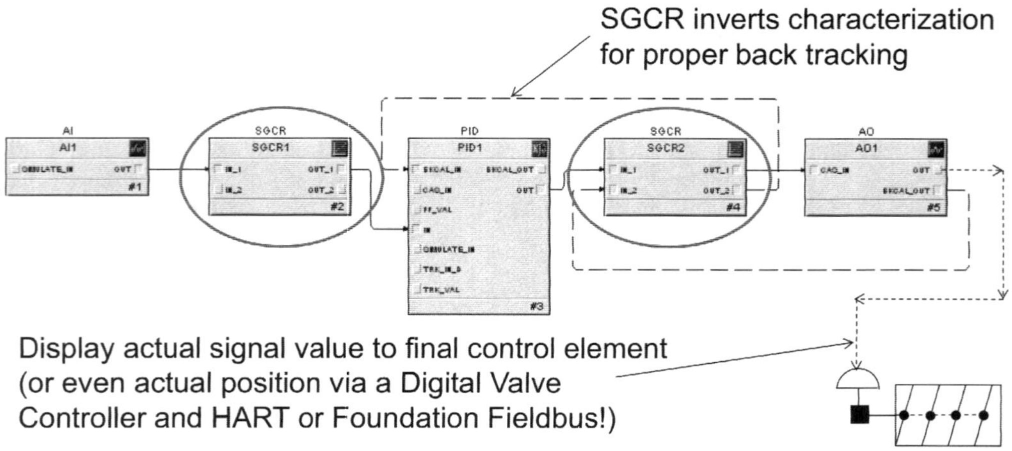

Signal Characterization – New Technology Solution #2

- Adaptive Control
  - Adjust tuning to compensate for final control element characteristics

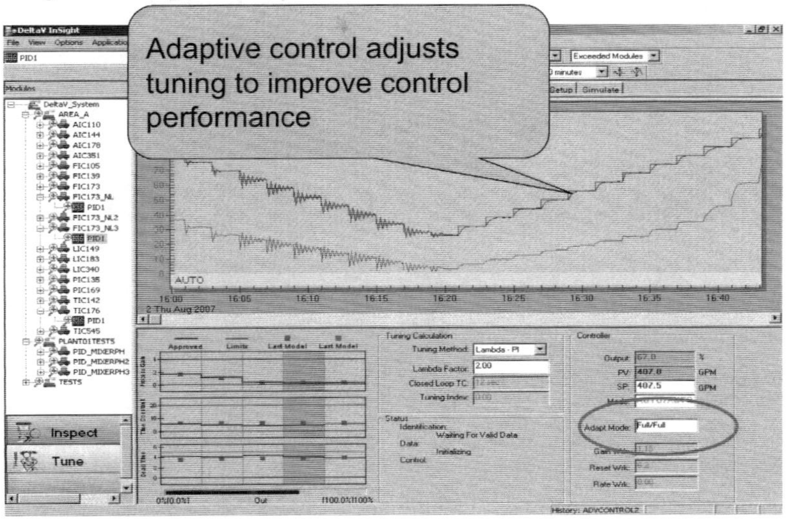

Summary Signal Characterization

- Newer Technology Signal Characterization function block
  - Better resolution (21 pairs of coordinates)
  - Self documenting (exists in control system)
  - No calibration required vs. 'external black box'
  - Provides "inverse" characterization for analog output characterizations
  - Reduce confusion: display signal being sent to final control element
- Applying Adaptive Tuning may provide a superior solution for AO characterization of final control element

Agenda

- Introduction
- Examples – Best Practices & Lessons Learned
  - Split Range Duty
  - Smith Predictor
  - Gain Scheduling/Adaptive Control
  - Controller Characterization
- Summary

Summary

- Practical application of control features that deal with common challenges
- Highlighted best practices & lessons learned
  - Split Range Duty
  - Smith Predictor
  - Gain Scheduling/Adaptive Control
  - Signal Characterization
- Do look at new functionality available when modernizing
  - Allows jump start to process optimization
  - Leads to improved process performance
  - Easier to implement and maintain over the life investment

Contacts & Questions?

James.Beall@Emerson.com
903-235-7935

Laurie.Ben@Emerson.com
512-834-7237

Paralleled Double-Effect Distillation: Simulative Case Studies

*Pei Wang, Leilei Dai, Ming Shi , Fan Wang, Hong Gao, Jinsheng Sun**
School of Chemical Engineering and Technology, Tianjin University, No. 92 Wei Jin Road, Tianjin 300072, P. R. China

Abstract

The increasing scarcity and growing demand for energy have led to an increasingly attention to the energy-efficient methods. Double-effect distillation is a common method to achieve remarkable energy conservation, which has three kinds of typical configurations including the feed splitting (FS), the light split/forward (LSF) and the light split/reverse (LSR). In spite of a good pathway for energy saving, better options are expected to cut down energy expenditure by a bigger margin than double-effect distillation does.

In this paper, case studies of paralleled or combined double-effect distillation schemes are proposed on the basis of current double-effect distillation. In order to verify the feasibility of this method, the improved methanol distillation scheme and organosilicon distillation scheme are optimized under this principle as examples. Exergy losses and the energy consumptions are measured as a tool to compare the two prevalent schemes with their corresponding optimized counterparts, respectively. The results of the total exergy loss and the energy consumption of these schemes show that the optimized schemes can cut the corner of the exergy loss by 22.8% and 40.41% and the energy consumption by 15.61% and 44.28%, respectively, in comparison to their popular cousins.

Keywords: double-effect distillation, paralleled double-effect distillation, methanol distillation, organosilicon monomer distillation, distillation process simulation, energy saving, exergy analysis

1. Introduction

Distillation, in chemical process and related industries, is by far the most widely applied for the separation of liquid mixture. In spite of its well-known benefit and widespread application, the major drawback is its high energy consumption. In classical distillation process, heat is transferred to the reboiler of distillation column. Along with the distillation column to separate the mixture, a large number of heat are directly discharge into the surrounding environment and cooling water. Araujo AB [1] has done some research to achieve that distillation has a relatively low thermodynamic efficiency and it requires the input of high quality energy in the reboiler to push the separation of liquid mixture. Therefore, the focus of energy conservation is to consider how to recover or reuse the discharged heat, or how to reduce the quantity of transferred heat to the column. In the early 20th century, many scholars have devoted to the study of saving energy in distillation, and now most of correlative techniques are widely used in distillation process [2,3]. These energy saving methods involve multi-effect distillation, heat pump distillation, heat integrated distillation and etc [3-8]. But without extra rotating machines by electricity or other power producers, multi-effect distillation has been more favored in research and application [9].

Many researches confirmed multi-effect distillation to be one of the effective methods to reduce the energy consumption [10-12]. The energy saving effect of those schemes used multi-effect distillation is associated with the number of the effects and the characteristics of separation system. Generally speaking, those schemes can offer energy savings around 30% with respect to the original or classical multicomponent distillation scheme [23, 27, 28]. Considering the interaction between investment of plants and energy saving effect, the double-effect distillation has been the most commonly used in not only research but also field application.

The use of the double-effect distillation technology for optimizing the conventional distillation schemes can cut a large amount of the energy requirement in comparison with the correspondingly prevalent schemes [13]. Although the system is very difficult to control in the sense that it is nonlinear, multivariable and interacting [14]. More and more published studies [15-19] have been focused on further questing the way of energy saving on the basis of the double-effect distillation schemes. However, the overhead vapor of one high pressured column only can integrate with the reboiler of one low pressured column in those published studies. In view of this, in this paper, the paralleled or combined double-effect distillation is proposed to broaden the thought for the application of the double-effect distillation, in which the overhead vapor from one of the high pressured column is divided into two or more streams to successively integrate with the reboiler of the other columns [9].

2. *Theory*

2.1 Double-effect or heat integrated distillation system

A double-effect distillation system is made up of two adjacent distillation columns running in different pressure, which can lead to a temperature difference between the top of the high pressured column and the bottom of the low pressured column. Through the thermodynamic connection, the reboiler of the low pressure column plays a role of the condenser for the high pressure column. According to the direction of the mass flow and the heat integration, the double-effect distillation can fall into three kinds of configuration: the feed splitting (FS), the light split/forward (LSF) and the light split/reverse (LSR), as showed in Figure. 1.

One of the configurations with a high thermodynamic efficiency is the feed splitting (FS), in which the feed is divided roughly equal between the two columns running in different pressure, and the products can be obtained from both of the top and bottom of the two columns, as illustrated in figure. 1 (a). The overhead vapor form the high pressured column is used to heat the reboiler from the low pressured column. There is only energy exchange between the two columns, so the heat duties of the feed splitting (FS) are approximately equal to the heat duty for single column with the same conditions divided by the number of effect.

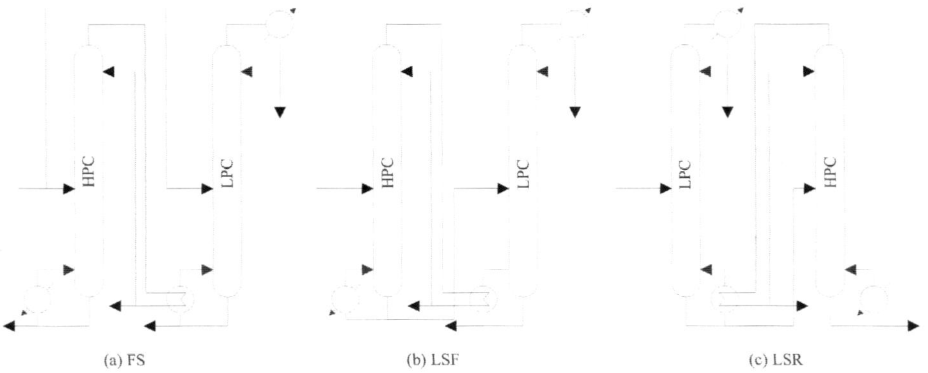

Fig. 1. Three configurations of double-effect distillation

Figure. 1 (b) and (c) respectively shows the configuration of the light split/forward heat integration (LSF) and the light split/reverse heat integration (LSR). Different from the feed splitting (FS), the total feed stream was sent to one column in the other two configurations. In the light split/forward (LSF), the total feed is pumped to the high-pressure column (HPC) and the bottom stream of HPC flows into the low-pressure column (LPC). In this way, all of the heavy components are from LPC and the light components from the top both of the two columns. Figure. 1 (b) indicates the direction of mass flow and heat integration is the same. The last configuration is the light split/reverse (LSR), as shown in figure. 1 (c), all of the feed is sent to the low-pressure column (LPC). It can remove the two kinds of light components from the top both of the two columns and one kind of heavy components from the bottom of HPC. Moreover, the direction of heat integration is in the reverse with the mass flow. These two kinds of configurations have not only energy exchange but also substantial exchange, therefore, when the operating parameters of one column are changed, the other column will also change.

Theoretically, the three kinds of configurations are empirically 30% less energy consumption than the single column distillation. As for choosing what kind of the configurations of the double-effect distillation, it depends on the properties of the multi-component mixture, the pressure of saturated steam, the temperature of cooling water, structure parameters of existing distillation column and so on.

2.2 Exergy analysis of the distillation process

According to the First Law of Thermodynamics, also called the Law of Energetic Balances, no matter what happens, the total amount of energy has not yet changed. However, when the various types of energy generate the mutual conversion, the quality of the energy is always decreased. The separation process in distillation is irreversible. In order to separate the mixture in distillation column, it needs the driving force which must consume a certain amount of external work, by this means, the decline of the quality of energy is inevitable. Therefore, we should choose a reasonable driving force in distillation process to avoid unnecessary loss of the quality of energy. It defined the concept of exergy (E_x) to compare the quality of the different types of energy. Exergy, which is a measure for evaluating the extent of available energy or the quality of energy, is based on the Second Law of Thermodynamics. Exergy analysis reveals the essence of the energy, and can be used as an important tool to evaluate the effect of energy saving. And it usually combines with process optimization studies, and as its guidance, to identify cost saving opportunities and amend the existing scheme.

Szargut [29] has summarized a comprehensive definition of the exergy, that is: 'Exergy is the amount of work obtainable when some matter is brought to a state thermodynamic equilibrium with the common components of the natural surroundings by means of reversible processes, involving interaction only with the above mentioned components of nature'. The state of one stream recorded as 1, according to the definition of exergy, the exergy of the stream can be expressed as:

$$E_x = (H_1 - H_0) - T_0(S_1 - S_0) \quad (2)$$

where, H_1 and S_1 is the enthalpy and the entropy of the stream 1, and T_0 is the temperature of the surrounding and H_0 and S_0 represent the enthalpy and the entropy of the surrounding.

The state of another stream recorded as 2, when the system stream change from state 1 (T_1, P_1) to state 2 (T_2, P_2), the variation of exergy (ΔE_x) is given by

$$\Delta E_x = E_{x,1} - E_{x,2} = \Delta H - T_0 \Delta S \quad (3)$$

In the conventional distillation process, it must input the heat or energy to drive the occurrence of separation process. The bottom of the column input the heat (Q_R) at this temperature (T_R), and the top of the column was sent to the heat (Q_C) at this temperature (T_C). The net work consumption (W) in distillation is

$$W = Q_R(1 - \frac{T_0}{T_R}) - Q_C(1 - \frac{T_0}{T_C}) \quad (4)$$

When the enthalpy difference between the product and the feed is very small, $Q_R=Q_C=Q$ can be approximately gotten, then Equation (4) is rewritten to be:

$$W = QT_0(\frac{1}{T_C} - \frac{1}{T_R}) \quad (5)$$

Then, the total exergy loss (Ex_{loss}) of the distillation column can be calculated by

$$Ex_{Loss} = W - \Delta E_x \quad (6)$$

3. *Prototype case: the prevalent distillation schemes*

3.1 The improved four-column methanol distillation scheme

In the methanol industry, it has gradually appeared two-column, three-column, four-column and five-column distillation scheme [20-23]. With the increase of the number of distillation column, the corresponding distillation scheme can not only increase the efficiency of the process, but also cut the energy consumption. At present, the four-column scheme has a dominant status to produce the purified methanol. However, the improved four-column distillation scheme can cut some more energy requirement with respect to the traditional four-column distillation scheme. The improved four-column distillation scheme is seen as the comparative reference in the paper.

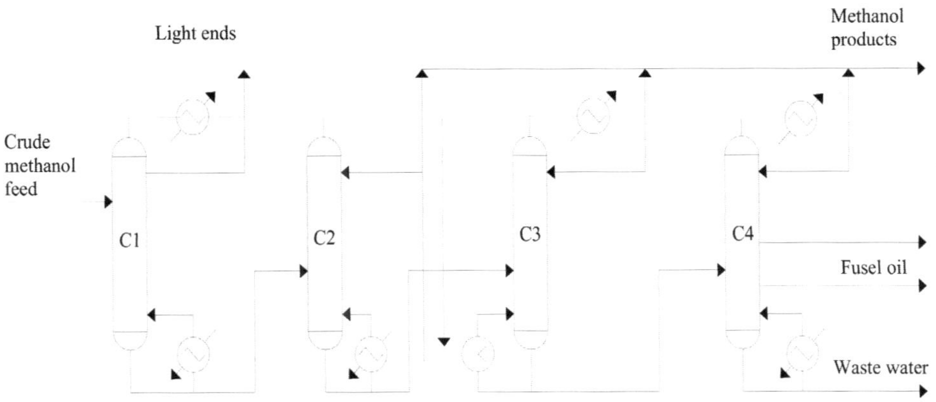

Fig. 2. The improved four-column double-effect methanol distillation scheme (C1, light ends column; C2, pressured column; C3, atmospheric column; C4, water column)

In the improved four-column distillation scheme, as shown in figure. 2 [24], the crude methanol flows into the light ends column (C1) in 87.5 ℃ for removing the light ends and the remnants of dissolved gas from the top of C1. Besides, in order to raise the methanol recovery rates, the light ends column (C1) designs a two-stage condenser system. The crude methanol was distillated successively from the pressured column (C2) and the atmospheric column (C3), and most of the methanol products can be received at the top both of C2 and C3. The bottom stream of C1, C2 and C3 is pumped respectively as the feed into C2, C3 and the water column (C4). Residual methanol, fusel oil and waste water are extracted from the corresponding position of the top, side stream and the bottom of the water column (C4).

In the improved scheme, the operation pressure of C2 is 0.9MPa, and the other three columns run under the atmospheric pressure. Due to the different of operating pressure, the overhead vapor of C2 can undertake thermal coupling connection with the reboiler of C3.

3.2 The current organosilicon monomer distillation scheme

According to the survey, there are several kinds of organosilicon distillation schemes. However, the 9-column organosilicon distillation scheme [9], in China, has an important role in the organosilicon industrial, as illustrated in Figure. 3, it is used as the base case.

Figure. 3 shows that the crude monomer is delivered to the heavies removal column (C1') and the bottom stream of C1' has a lot of the unsteady substances of high boiling points. These unsteady substances of high boiling points are pumped to the high boiling point column (C9') as the feed, and the products from the top of C9' continue to serve as the crude monomer by circumfluence. By this way, it can accomplish the separation between the unsteady substances of high boiling points and the useful substances at a relatively low temperature in the high boiling point column (C9') and avoid the unsteady substances of high boiling points from thermal cracking to generate many unknown substances which would greatly affect the subsequent separation of product. The low removal column (C2') provides trimethylchlorosilane (Me3), methyldichlorosilane (MeH), and other light components for lights removal column (C3'), MeH column (C4'), azeotrope removal column (C5') and Me3 column (C6'), which can finally achieve two products (MeH and Me3) and three wastes. It also provides methyltrichlorosilane (Me1) and dimethylchlorosilane (Me2) and other heavy components for Me1 column (C7') and Me2 column (C8'), which can finally get two products (Me1 and Me2).

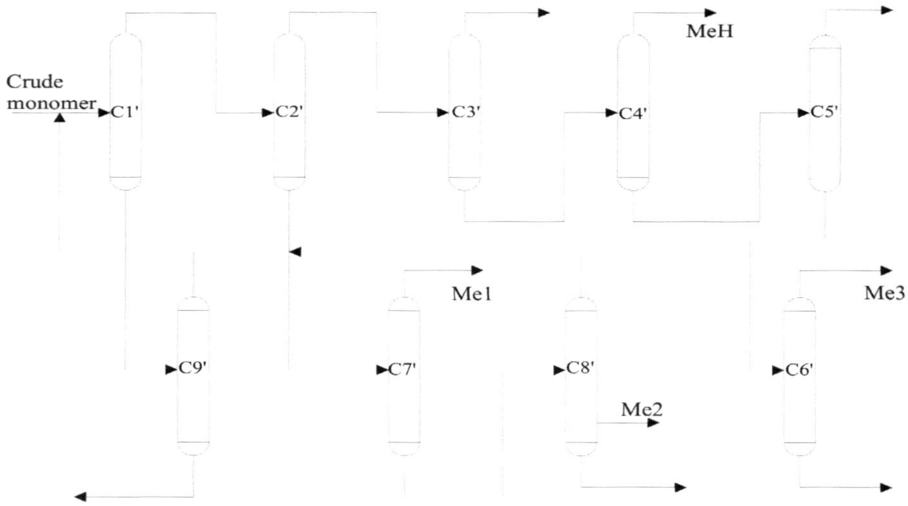

Fig. 3. The prevailing 9-column distillation scheme (C1', heavies removal column; C2', low removal column; C3', lights removal column; C4', MeH column; C5', azeotrope removal column; C6', Me3 column; C7', Me1 column; C8', Me2 column; C9', high boiling point column)

In this scheme, the operating pressure of all nine columns is relatively low, and the highest operating pressure is 260 kPa in the lights removal column (C3'). Although it has a different operating pressure between C3' and the other columns, but the temperature difference is not enough to make them use double-effect or heat integrated distillation.

4. *Simulation and analysis both of the prevalent schemes and the optimized schemes*

4.1 The novel five-column heat integrated methanol distillation scheme

Although the energy saving of the improved four-column methanol distillation scheme can reach 10% compared with the traditional four-column scheme, it still exist several shortages, which will be illustrated by the results of computer simulation to the improved four-column methanol distillation scheme. The flow rate of the crude methanol is 239230 kg/hr, and the temperature is 87.5 ℃, and the composition of the crude methanol is exhibited in table 1. The inlet and outlet temperature of the cooling water are 30 ℃ and 40 ℃.The methanol content is specified as 99.99 wt% in the methanol products and 10ppm in the wastewater of the water column. Table 2 shows the operating parameters of each column in the improved four-column scheme.

Table 1. Crude methanol feed data

Component	Mass fraction%	Component	Mass fraction%
H_2	0.0100	Pentanol	0.0338
N_2	0.0710	Formic acid	0.0030
Methane	0.0100	Acetic acid	0.0030
CO	0.0360	Methyl formate	0.0092
CO_2	0.6640	Acetone	0.0193
Methanol	95.0000	Methyl ethyl ketone	0.0030
Dimethyl ether	0.0309	Diisopropylether	0.0056
Trimethylamine	0.0010	Water	3.7100
Ethanol	0.0900	i-Pentane	0.0562
i-Propanol	0.0024	n-Pentane	0.0169
Propanol	0.0264	n-Hexane	0.0487
i-Butanol	0.0338	n-Heptane	0.0009
Butanol	0.0185	n-Octane	0.0050
3-Pentanol	0.1407	n-Nonane	0.0050

Table 2. The operating parameters of improved four-column scheme

The operating parameters	C1	C2	C3	C4
Top pressure (kPa)	150	900	110	110
Top temperature (℃)	81.0	132.6	67.7	67.7
Bottom temperature (℃)	83.8	139.1	96.5	104.5
Reflux ratio	0.58	2.44	1.68	11.57
Number of theoretical stage	33	52	58	61

The results of simulation indicate that the hot utility is still in a high level because the reboilers of C1, C2 and (C4) still need the saturated steam to transmit the heat. Table 2 shows the reflux ratio of C4 is 11.57 and this large reflux ratio can cause the increasing of the energy requirement of the water column (C4). Besides, according to the flowsheet of this scheme, the diameter of column is so large that the emergence of many urgent cases.

In view of above these defects in the improved four-column scheme, the novel five-column heat integrated methanol distillation scheme (shown as Fig. 4) is designed in this paper to reduce the impact of these factors as far as possible by using the paralleled or combined double-effect distillation. Juntao Zhang [23] and Jinsheng Sun [25] are also proposed a five-column methanol distillation scheme by using triple-effect distillation and double-effect distillation respectively, and the effect of energy saving of the two schemes is outstanding.

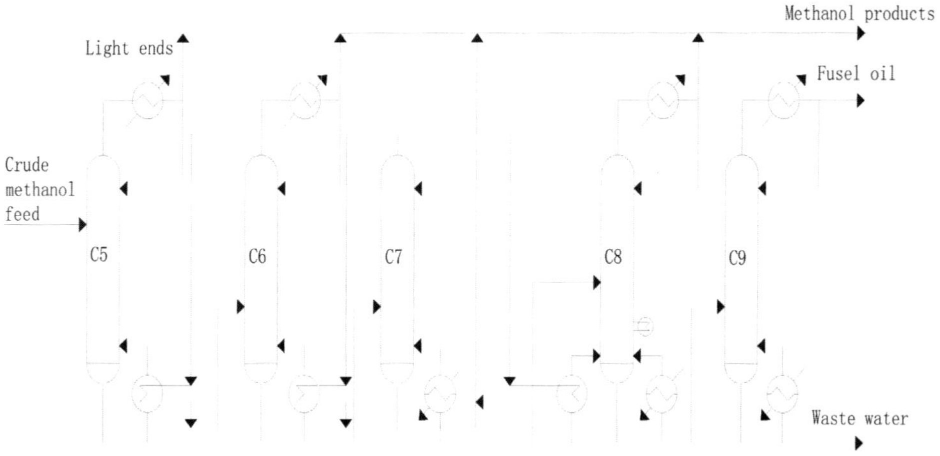

Fig. 4. The five-column heat integrated methanol distillation scheme (C5, light ends column; C6, prepositive atmospheric column; C7, pressured column; C8, postpositive atmospheric column; C9, water column)

According to the principle of the paralleled double-effect distillation, the four columns are unable to constitute the heat integration network in the improved four-column scheme. Through repeated verification, the prepositive atmospheric column (C6) is added after the light ends column (C5), meanwhile, the operating pressure of the pressured column (C7) is reduced to increase its throughput. In this way, the overhead vapor of C7 is split into three streams to respectively integrate with the light ends column (C5), the prepositive atmospheric column (C6) and the postpositive atmospheric column (C8). Due to use the paralleled double-effect distillation, the duties of the hot utility have decreased in a great margin. However, there are still part of heat from the background process is absorbed by the cooling water. So an inter-reboiler with a low temperature level is designed in the postpositive atmospheric column (C8) and it can share the heat with the background process streams, besides, we have proved the feasibility by using the column grand composite curve analysis of columns. Figure. 4 shows that the product of the top of the water column (C9) is fusel oil instead of purified methanol, which can not only reduce the reflux ratio of C9, but also make the diameter of column smaller to reduce the difficulty of operation.

Due to produce the fusel oil instead of purified methanol in the water column (C9) in the novel five-column distillation scheme, the methanol content of the feed for the water column have to limit at 0.016 wt%. In the five-column scheme, the operating pressure and the number of theoretical stage for each column are determined through the existing process, the onsite data and the requirement of the heat coupling. The proportion of the methanol products from the C6, C7 and C8 is determined by their own double-effect or heat integrated distillation system. The simulation results of the novel five-column scheme are listed in table 3.

Table 3. The operating parameters of five-column scheme

Column	C5	C6	C7	C8	C9
Top pressure (kPa)	160	110	400	110	110
Top temperature (℃)	83	66.6	104	66.6	86.5
Bottom temperature (℃)	85.6	84	118.8	104.5	103.2
Reflux ratio	1.46	1.49	2.23	1.87	4
Number of theoretical stage	33	54	54	65	29

4.2 Energy consumption and exergy loss analyses for the two methanol schemes

In the improved four-column scheme, the pressured column and the atmospheric column constitute the distillation system of heat coupling, there are three columns (C1, C2, C4) need the hot utility, and the entire duty of reboiler is 145.4×10^3 kw. As described above, the light ends column (C5) and the prepositive atmospheric column (C6) don't need the saturated steam in the five-column scheme and the total duty of reboiler is reduced to 122.7×10^3 kw. The detailed data of the reboiler duty as shown in table 4.

Table 4. The duty of reboiler in the two schemes

Column	C1	C2	C4	C7	C8	C9
Duty of reboiler (10^4 kw)	3.30	10.04	1.20	8.97	2.86	0.44

Obviously, the novel five-column scheme is much better than the improved four-column scheme in energy saving and the energy consumption of the novel scheme can be reduced by 15.61%. However, the recovery rate of methanol of the novel scheme is approximately equal with the improved four-column scheme.

Only the energy consumption analysis is not revealed the energy saving essence of the novel five-column scheme. Exergy analyses are used to further compare the two schemes, and the exergy flow rates are calculated through using the computer simulation data. The activity coefficient of the liquid is calculated using the Alcohol Package thermodynamic model. The calculation of the net work consumption of distillation column should pay more attention to the temperature of the cooling water. When the cooling water exchanges heat with other streams, the temperature of the cooling water is momentarily changed, so it is replaced the temperature of cooling water with the arithmetic average of inlet and outlet temperature of cooling water. Figure. 5 show the overall exergy loss of each column in the two methanol distillation schemes, in addition to the overall exergy loss of C6. Due to optimize the four-column scheme, the prepositive atmospheric column (C6) is added in the transformed scheme, and the overall exergy loss of C6 is 9157.6 kw. The exergy loss of each column, besides the pressured column (C7), in the novel five-column scheme is obviously less than their counterparts in the improved four-column scheme.

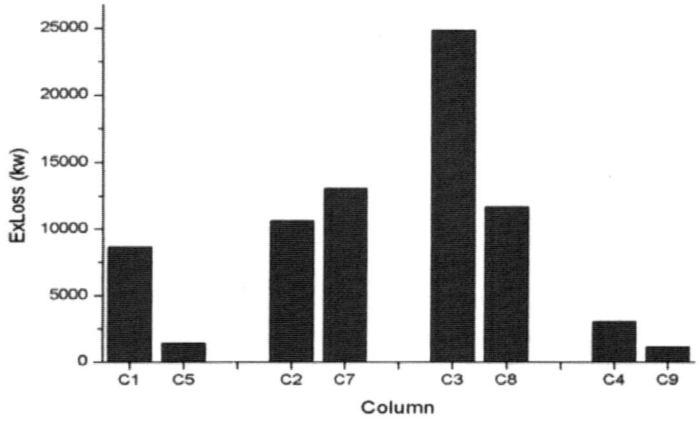

Fig. 5. The overall exergy loss analysis of each column in the two schemes

The total exergy loss of the improved four-column scheme and the transformed scheme is, respectively, 47100.1 kw and 36355.1 kw. The total exergy loss of the optimized scheme is dropped by 22.8%. Through the comparison between the exergy loss and energy consumption of the two schemes, it is indicated that the novel five-column scheme should be applied to produce the purified methanol in industry.

4.3 The novel heat coupling organosilicon distillation scheme

The primary source of the crude organosilicon monomers is the catalytic reaction products between chloromethane and silicon at high temperature and the table 5 displays the normal boiling points and mass percentage of the main components of the crude organosilicon monomers. Methyltrichlorosilane (Me1), dimethylchlorosilane (Me2), trimethylchlorosilane (Me3) and methyldichlorosilane (MeH) are the most valuable components in the crude organosilicon monomers. The purification requirements on organosilicon of these materials are shown in table 6.

Table 5. The crude organosilicon monomer feed data

Component	Normal boiling point at 1 atm/℃	Mass percentage %
LB		1
Methyldichlorosilane	41.8	4
Silicon tetrachloride	57.6	0.5
Trimethylchlorosilane	57.9	2.5
Methyltrichlorosilane	66.4	11
Dimethyldichlorosilane	70.2	80
HB		1
Temperature (℃)	40	
Pressure（kPa）	100	
Flow rate (kg/h)	26200	

Table 6. Purification demand on organosilicon

Monomer	Purification demand
MeH	99.5%
Me3	99.5%
Me1	99.5%
Me2	99.95%

The conventional nine-column scheme is simulated with the aid of existing process and onsite data to achieve the operating parameters of each column, as illustrated in table 7. The bottom temperature of the heavies removal column (C1') and the high boiling point column (C9') are obviously much bigger than that of the other seven columns in table 7. According to the principle of the double-effect distillation, the top temperature of the high pressured column is relatively high in comparison with the bottom temperature of the low pressured column, and the high pressured column can integrate with the low pressured column. Although the top temperature of C1' is less than the bottom temperature of the other seven columns, the overhead vapor of C1' still has a great opportunity to conduct heat with the other seven columns with the increasing of its top temperature.

Table 7. The simulation results of prevailing 9-column

Column	C1'	C2'	C3'	C4'	C5'	C6'	C7'	C8'	C9'
Top temperature (℃)	70.8	65.3	57.3	66.7	72.1	70.6	71.2	74.9	72.7
Bottom temperature (℃)	124.1	89.1	80.4	85.2	74.9	75.4	75.3	75.3	126.5
Reboiler duty (MW)	11.67	5.61	1.13	0.51	1.19	0.46	12.27	16.08	0.02
Condenser duty (MW)	-12.47	-5.53	-1.08	-0.53	-1.20	-0.47	-12.69	-16.12	-0.02
Top pressure (kPa)	110	180	260	220	170	150	110	110	110

According to the way of the paralleled double-effect distillation, the increasing operating pressure in the heavies removal column (C1') can make the top temperature of C1' ascending to integrate with the other seven columns. The idea is still in study to overcome some difficulties. However, we have got another feasible method, in which the operating pressure of C1' and the Me1 column (C7') is raised to 360 kPa and 280 kPa, respectively. In this way, the top temperature of C1' and C7' is increased to ensure they can integrate with the other columns, besides the high boiling point column, and the parallel double-effect or heat integrated organosilicon monomer distillation scheme is designed as shown in figure. 6 [26]. The feasibility of this heat coupling scheme has confirmed through three x-y diagrams including the heavies component and Me1, the heavies component and Me2, Me1 and Me2 [9].

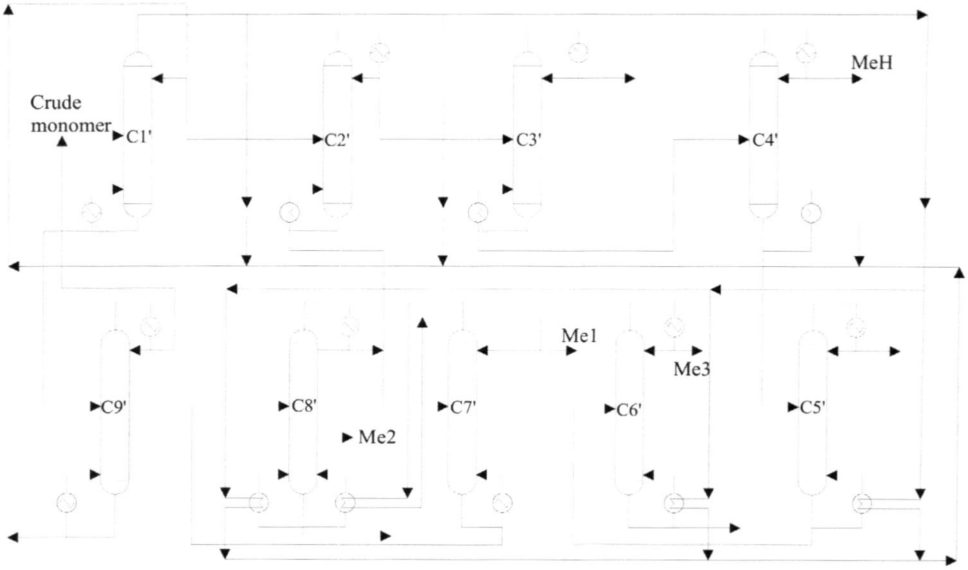

Fig. 6. The heat coupling 9-column scheme (C1', heavies removal column; C2', low removal column; C3', lights removal column; C4', MeH column; C5', azeotrope removal column; C6', Me3 column; C7', Me1 column; C8', Me2 column; C9', high boiling point column)

Figure. 6 shows the heavies removal column (C1') make the heat coupling with the low removal column (C2'), the lights removal column (C3'), the MeH column (C4'), the azeotrope removal column (C5'), the Me3 column (C6') and the Me2 column (C8'), successively. In order to gain a better energy saving effect, C8' is designed with two reboilers, the one integrated with C1' and the other integrated with the Me1 column (C7'). The operating parameters and the duty of reboiler and condenser of each column in the heat integrated 9-column are enumerated in table 8.

Table 8. The simulation results of heat integrated 9-column

Column	C1'	C2'	C3'	C4'	C5'	C6'	C7'	C8'	C9'
Top temperature (℃)	114.3	65.3	57.3	66.7	72.1	70.6	105.1	74.9	72.7
Bottom temperature (℃)	124.9	89.9	78.8	84.5	75.4	75.4	110.9	76.6	121.4
Reboiler duty (MW)	13.06	5.62	1.13	0.51	1.19	0.46	13.83	14.12	0.38
Condenser duty (MW)	-13.72	-5.54	-1.08	-0.53	-1.20	-0.47	-14.02	-14.43	-0.41
Top pressure (kPa)	360	180	260	220	170	150	280	110	110

4.4 Energy consumption and exergy loss analyses for the two organosilicon schemes

Figure. 4 indicate that each column of the prevalent nine-column scheme is not integrated with the other columns and the reboilers of all of these columns are required the saturated steam (600 kPa). According to the data from the table 7, the total duty of reboiler is 48.94 MW for the prevailing 9-column scheme. In the new nine-column scheme, C1' integrate with C2', C3', C4', C5', C6' and C8' successively and C7' integrate with C8', so C2', C3', C4', C5', C6' and C8' don't consume the saturated steam. The total duty of reboiler is 27.27 MW. Obviously, the heat integrated 9-column scheme has an absolute advantage over the prevailing 9-column scheme in energy saving, and the energy savings can reach 44.28%.

As the same with the methanol distillation scheme, the inlet and outlet temperature of cooling water are 30 ℃ and 40 ℃, and the total exergy loss is calculated through the results of computer simulation of the two schemes. The overall exergy loss comparison of each column in the two schemes is illustrated in figure. 7. In the normal scheme, the throughput of C9' is so small that the overall exergy loss of C9' is very near to zero, so it only have the exergy loss of C9' of the new scheme in figure. 7. The overall exergy loss of each column in the optimized scheme is declined in comparison with that of the corresponding column in the normal scheme except the high boiling point column (C9'). Because the crude organosilicon monomers contains some unstable components, in order to avoid these components happening thermal cracking in the heavies removal column (C1') in the novel scheme, more liquid from the bottom of C1' is pumped to C9' as the feed. That can cause the ascent of the throughput and the overall exergy loss of C9'.

The total exergy loss is 14035.5 kw and 8363.2 kw, correspondingly, in the conventional scheme and the novel scheme, and the exergy loss of the new scheme can be reduced by 40.41%. Through the above analysis and comparison, the heat integrated nine-column organosilicon distillation scheme should be a substitute for the prevalent nine-column scheme.

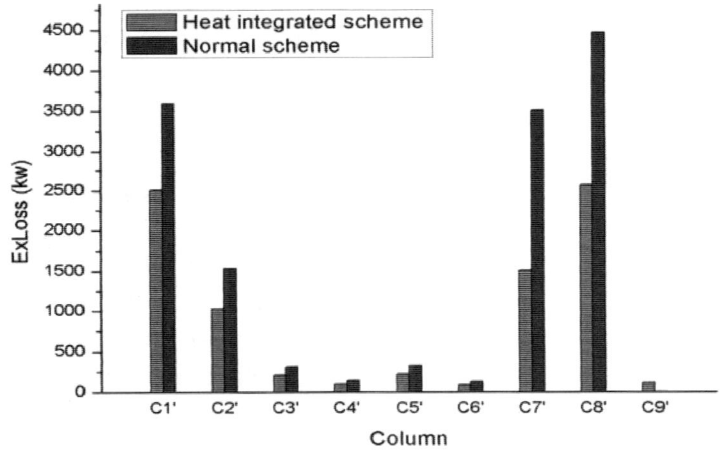

Fig. 7. The exergy loss comparison between the heat integrated scheme and the normal scheme

5. *Conclusions*

Paralleled double-effect or heat integrated distillation for energy saving is proposed, exampled by heat integrated five-column methanol distillation scheme and the heat coupling 9-column organosilicon monomer distillation scheme, with the aid of computer simulation. Besides, the optimized methanol distillation scheme and the transformed organosilicon scheme are comprehensively analyzed through the energy consumption and the exergy loss in comparison with the correspondingly prevalent scheme.

Adding a prepositive atmospheric column and decreasing the operating pressure of the pressured column in the novel five-column scheme can make the pressured column to simultaneously integrate with the light ends column, the prepositive atmospheric column and the postpositive atmospheric column. Besides, an inter-reboiler is added to integrate with the background process streams. The energy consumption and the exergy loss of the new scheme are 15.61% and 22.8% lower than the improved scheme.

Increasing the operating pressure of the heavies removal column and the Me1 column can make them to integrate with the other columns, in addition to the high boiling point column, to constitute a paralleled double-effect distillation system in the optimized nine-column scheme. The energy consumption and the exergy loss of the novel scheme are 44.28% and 40.41% less than the normal scheme.

The two cases expound that the key of paralleled double-effect distillation process lies in the reasonable adjustment of the energy distribution, the operating parameters and the configuration of some columns. It not only reduces the quantity of energy consumption, but also avoids the decrease of the quality of energy. For factories with a large numbers of distillation schemes of high energy requirement, the paralleled schemes lead to improve economic benefit.

References

[1] Araujo AB, Brito RP, Vasconcelos LS, "Exergetic analysis of distillation processes-a case study," Energy 2007;32:1185-93.

[2] Michele Mascia, Francesca Ferrara, Annalisa Vacca, Giuseppe Tola, Massimiliano Errico, "Design of heat integrated distillation systems for a light ends separation plant," Applied Thermal Engineering 2007;27:1205-1211.

[3] Amiya K. Jana, "Heat integrated distillation operation," Applied Energy 2010;87:1477-1494.

[4] Hilde K. Engelien, Sigurd Skogestad, "Multi-effect distillation applied to an industrial case study," Chemical Engineering and Processing 2005;44:819-826.

[5] James G. Gebbie, Michael K. Jensen, Piotr A. Domanski, "Experimental transient performance of a heat pump equipped with a distillation column," International Journal of Refrigeration 2007;30:499-505.

[6] Massimiliano Errico,Giuseppe Tola, Michele Mascia, "Energy saving in a crude distillation unit by a preflash implementation," Applied Thermal Engineering2009;29:1642-1647.

[7] Massimiliano Erricoa, Giuseppe Tola, Ben-Guang Rong, Daniele Demurtas, Ilkka Turunen, "Energy saving and capital cost evaluation in distillation column sequences with a divided wall column," chemical engineering research and design 2009;87:1649-1657.

[8] Ivar J. Halvorsen, Sigurd Skogestad, "Energy efficient distillation," Journal of Natural Gas Science and Engineering 2011;3:571-580.

[9] Jinsheng Sun, Fan Wang, Tingting Ma, Hong Gao, Yanzhen Liu, Fang Cai, "Exergy analysis of a parallel double-effect organosilicon monomer distillation scheme," Energy 2012;47:498-504.

[10] Hilde K. Engelien, Sigurd Skogestad, "Multi-effect distillation applied to an industrial case study," Chemical Engineering and Processing 2005;44:819-826.

[11] Kazuo Matsuda, Kenichi Kawazuishi, Yasuki Kansha, Chihiro Fushimi, Masaki Nagao, Hiroshi Kunikiyo, Fusao Masuda, Atsushi Tsutsumi, "Advanced energy saving in distillation process with self-heat recuperation technology," Energy 2011;36:4640-4645.

[12] M.A. Sharaf, A.S. Nafey, Lourdes García-Rodríguez, "Exergy and thermo-economic analyses of a combined solar organic cycle with multi-effect distillation (MED) desalination process," Desalination2011;272:135-147.

[13] Weiwen Wang, Wenjian Wang, "Energy Saving in Methylchlorosilane Distillation," International Conference on Computer Distributed Control and Intelligent Enviromental Monitoring 2012:290-293.

[14] Myungwan Han, Sunwon Park, "Multivariable control of double-effect distillation

configurations," J. Proc. Cont. 1996:247-253.

[15] Hilde K. Engelien, Sigurd Skogestad, "Multi-effect distillation applied to an industrial case study," Chemical Engineering and Processing2005;44:819-826.

[16] Debadrita Maiti, Amiya K. Jana, Amar Nath Samanta, "A novel heat integrated batch distillation scheme," Applied Energy2011;88:5221-5225.

[17] Anton A. Kiss, Servando J. Flores Landaeta, Carlos A. Infante Ferreira, "Towards energy efficient distillation technologies-Making the right choice," Energy2012;47:531-542.

[18] Diego C. Alarcón-Padilla, Lourdes García-Rodríguez, Julián Blanco-Gálvez, "Experimental assessment of connection of an absorption heat pump to a multi-effect distillation unit," Desalination2010;250:500-505.

[19] Diego C. Alarcón-Padilla, Lourdes García-Rodríguez, Julián Blanco-Gálvez, "Design recommendations for a multi-effect distillation plant connected to a double-effect absorption heat pump: A solar desalination case study," Desalination2010;262:11-14.

[20] Fiedler E, Grossman G. Methanol, "Ullmann's Encyclopedia of Industrial Chemistry," Weinheim : Wiley-VCH Verlag/Gmbh & Co; 2002.

[21] Meyers RA, "Handbook of synfuels Technology," New York : McGraw-Hill; 1984.

[22] Chu YZ, Qin LP, "Optimum design of methanol distillation process and columns," Chem Ind Eng Prog 2008;(27):1659-62.

[23] Juntao Zhang, Shengrong Liang, Xiao Feng, "A novel multi-effect methanol distillation process," Chem Eng Pro 2010;(49):1031-37.

[24] Jinsheng Sun, Fan Wang, Tingting Ma, Hong Gao, Peng Wu, Lili Liu, "Energy and exergy analysis of a five-column methanol distillation scheme," Energy2012;45:696-703.

[25] J. S. Sun, P. Wu, Y. H. Wang, H. Gao, L.L. Liu.CN Patent102039058 A. 2011.

[26] J. S. Sun, P. Wu, Y. H. Wang, Y. X. Su, H. Gao, J. Li, D. Yan. CN Patent 101798322 A. 2010.

[27] Michele Mascia, Francesca Ferrara, Annalisa Vacca, Giuseppe Tola, Massimiliano Errico, "Design of heat integrated distillation systems for a light ends separation plant," Applied Thermal Engineering 2007;27:1205-1211.

[28] Olga A. Flores, J. Carlos Ca´ rdenas, Salvador Herna´ ndez, Vicente Rico-Ramı´rez, "Thermodynamic Analysis of Thermally Coupled Distillation Sequences," Ind. Eng. Chem. Res. 2003;42:5940-5945.

[29] Szargut J, "Exergy Analysis of Thermal, Chemical and Metallurgical Processes," New York: Hemisphere Publishing Corporation;1988:55-152.

(45n) In Situ Transesterification of Jatropha Curcas Seed Particles with Alkaline Phase Transfer Catalyst

Subbarao Duvvuri[1], Shuhaimi Mahadzir[2] and Sintayehu Hailegiorgis[2], (1)Department of Chemical Engineering, Universiti Teknologi PETRONAS, Tronoh, Malaysia, (2)Chemical Engineering, Universiti Teknologi PETRONAS, Tronoh, Malaysia

Economic viability of biodiesel technology is at stake due to the high cost of edible oil feedstock and slow reaction rates due to limited solubility of oil in methanol. In situ transesterification of non edible oil seed particles can minimize the cost of feedstock. Microwave pretreatment of seeds can make oil molecules more reactive. Use of phase transfer catalysts can enhance the reaction rates. In the present work, these three concepts are utilized together to investigate alkaline in situ transesterification of microwave pretreated jatropha curcas seed particles in presence of benzyltrimethylammonium hydroxide (BTMAOH) as a phase transfer catalyst (PTC). It was observed that addition of PTC to alkaline catalyst has improved conversion and reduced reaction time. Microwave pretreatment of jatropha seed particles reduced reaction time drastically. The quality of biodiesel produced was within the limits of the international standard requirements of ATME D6751 and EN-14214. Thus, combined use of microwave pretreatment of seed particles, phase transfer catalysis in the presence of alkaline catalyst can increase the rate of in situ transesterification and product yield.

(45o) Catalyzed-Assisted Manufacture of Olefins (CAMOL): Updated for Use in Naphtha Service

Maximilian Walter, Petrochemicals, BASF SE, Ludwigshafen, Germany and Suzanne Rech, BASF Qtech, BASF Corporation, Iselin, NJ

Since the commercialization of the olefins plant, operators have searched for ways to be more competitive. One area always targeted includes keeping their furnaces in operation longer between decoking steps. Over the years, many anti-coking methods/technologies have become commercially available using different approaches to prevent coils from coking. One of the latest available technologies is the novel CAMOL™ (Catalyzed-assisted Manufacture of Olefins) Technology from BASF Qtech. As reported in the introduction of the technology to the EPC in 2008 & 2010, the CAMOL™technology differs from other available anti-coking technologies in that the technology not only eliminates the catalytic (or filamentous) coke make through a physical barrier, but also gasifies the thermal (or amorphous) coke for an even greater coking resistance. The coating process has generated a coating with almost identical thermo-mechanical properties as the base metal substrate (alloy). This allows bonding to the base metal similar to being welded with a minimum risk of spallation when going through thermal cycling. Commercially, the CAMOL Technology has been available since 2006 and is now in operation in furnaces of various designs.

With CAMOL™, operators have experienced extended run lengths, reduced energy requirement and GHG emissions, lower average TMT operating temperatures, reduced DMDS consumption, and less carburization. In addition, the technology has exhibited high operating temperature stability (>1130°C/ 2065°F).

At BASF SE in Ludwigshafen, Germany, the technology has been in operation since early 2010 in mostly naphtha service. After overcoming some initial challenges associated with a catalytic surface technology, such as reactivating the catalytic species after decoking, the coil is now successfully operating with run lengths significantly higher than uncoated coils. This paper will report on the results and learnings from the experiences gathered over the past three years along with the advancements made since the last report in 2010.

(45p) Extending Furnace Run Length and Coil Life by Optimizing Decoking

Vinod Mishra, Ingenero, Mumbai, India and **James Brigman**, Ingenero, Houston, TX

Abstract:

In the Ethylene manufacturing process, coke deposited in the radiant coils and Transfer Line Exchangers (TLEs) during normal operation is removed periodically by decoking the furnaces using a mixture of steam and air. There is a significant variation in operating parameters adopted by different plants for decoking. These are often either too conservative or aggressive, leading to production loss from increased off-stream time, TLE limitation and deterioration in coil life. This paper describes how the decoking procedure was modified based on furnace and TLE operating conditions and the decoking steps changed to obtain the best possible Ethylene yield, without adversely impacting coil life and TLE performance. This optimization has now been done in several major ethylene facilities and furnace run length has more than doubled and production has significantly increased.

SOLVING OPERATIONAL PROBLEMS IN A NAPHTHA SPLITTER

Flávio Weissheimer, GE Water & Process Technologies do Brasil Ltda, São Paulo/Brasil
Giorgio Franceschetti, Braskem, Santo André/Brasil
Rafael Henrique Camparin, GE Water & Process Technologies do Brasil Ltda, São Paulo/Brasil
Patrick Lucas, GE Water & Process Technologies, The Woodlands/US

Abstract

This paper sets out how the polymerization process of Reboilers of the (A-100) Crude Naphtha Fractionation Unit of Braskem UNIB 3 ABC can be controlled by applying polymerization inhibitors developed by GE Water & Process Technologies Ltda and Braskem´s engineering solutions.

These heat exchangers have a high degree of polymerization because of the content of olefinic compounds present in the feed composition which, when in contact with high temperatures in the heat exchange areas of reboilers, tend to polymerize. The joint solution of GE technology and Braskem engineering solutions applied to this system ensures regular operation of such equipment for periods of time greater than those hitherto recorded without manifesting loss of heat transfer efficiency.

In the 2012 shutdown while the opening of these reboilers was being inspected, it was observed that pipes had not been clogged nor was surface fouling present, which is a very positive state when compared to the historical records of previous campaigns.

The strategy adopted by GE/Braskem to ensure a longer campaign for these pieces of equipment was based on the quality control of feeding by adjusting dosages and monitoring performance using statistical analysis tools.

The use of statistical models based on Multivariable Regression Analysis (MRA) to evaluate performance, which is validated by inspecting reboilers internally, has been an efficient way to determine the degree to which the system has been fouled and also to schedule shut-downs for cleaning the mechanical parts.

Description of the Process

The (A-100) Naphta Fractionation Unit includes two distillation columns, the objective of which is to separate the light (C_6), medium (C_6 - C_8) and heavy (C_{9+}) cuts, included in the feed supply of crude naphtha.

The light cut and a portion of the heavy cut are sent to pyrolysis furnaces so as to produce ethylene. The medium cut is sent to the Hydrotreating and Catalytic Reforming Unit, with the objective of increasing the aromaticity of the system so as to produce BTX.

The A-100 processes 100 tons/h of crude naphtha, and approximately 35% of this current represents the average cut. Figure 1 below shows the schematic flow diagram of this fractionation process.

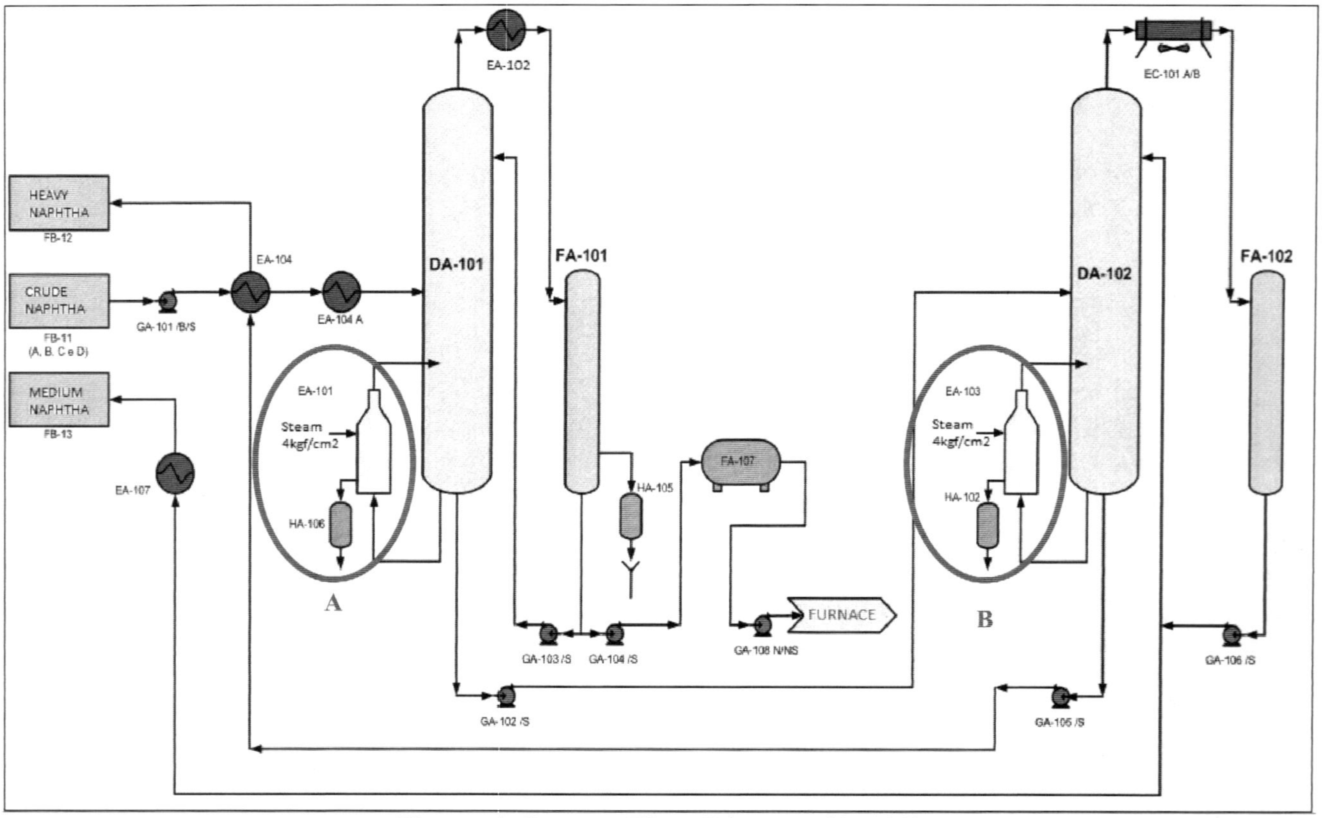

Figure 1. Process Flow Diagram (A-100)

The Reboilers, indicated by the letters A and B in Figure 1 above, operate with an injection of low steam (4kgf/cm2) to promote vaporization of the bottom stream of the A-100 distillation columns. These devices operate over a temperature range of about 150°C, which is sufficient to promote the polymerization of certain olefinic compounds present in the feed of crude naphtha.

Poor performance of these Reboilers can lead to production stoppages for maintenance and cleaning, which very often affects up to five days of production. The possibility of raising the times of campaigns and predicting the degree of fouling in downstream units helps to minimize production losses due to unscheduled stoppages.

This study will discuss the results of Reboiler A, which is the most critical equipment, with a view to synthesizing data, but gains in equipment B were also significant.

Evaluation System

Historical Record of the Campaign

Reboiler A was installed in 1972 but examination of the historical records did not reveal that the sheaf of tubes had ever been replaced. This exchanger has currently 11 tubes plugged into it.

The Campaign times of this equipment are shown in Table 1, with the last campaign being highlighted in blue so as to set it apart from the other ones. Table 1 gives a summary in the order of the timings of the campaigns, the average load processed in the period, the total load processed, the reason for operational stoppages etc.

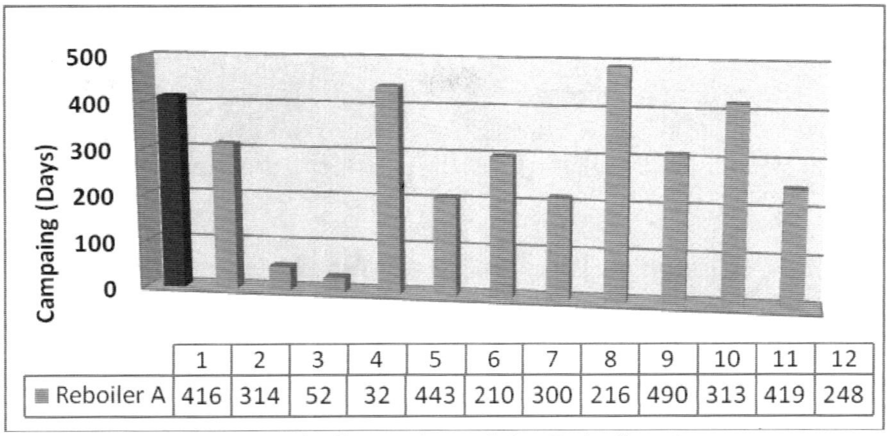

Figure 2. Campaign of the Reboiler A

Table 1. Historical Record of Reboiler A Campaigns

Date	Campaign *(days)*	Capacity *(ton/h)*	Capacity *(ton/campaign)*	Occurrence	Tubes Clogged *(%)*
13/06/12	416.0			Programmed Shutdown	0
17/03/11	314.0	93.1	701601.6	Heat exchange performance	30
03/05/10	52.0	90.8	113318.4	Stuck tube	-
11/03/10	32.0	71.1	54604.8	Heat exchange performance	50
09/02/10	443.0	72.0	765716.6	Heat exchange performance	20
16/11/08	210.0	98.1	494575.2	Heat exchange performance	-
16/04/08	300.0	104.0	748800.0	Programmed Shutdown	-
16/06/07	216.0	92.8	481075.2	Programmed Shutdown	-
10/11/06	490.0	-	-	Programmed Shutdown	-
30/06/05	313.0	-	-	Stuck tube	-
17/08/04	419.0	-	-	-	-
18/06/03	248.0	-	-	-	-
10/10/02	-	-	-	-	-

Conditions of Opening

In this review, a comparison is presented between two operational shut-downs of A-100, which occurred in March 2011 and May 2012, in order to observe the improvement in performance from the last campaign associated with the change in the GE Treatment strategy.

Campaign I: Production shutdown on 03/2011

Table 2 below summarizes the features observed during the inspection of the state of the inlet of this piece of this equipment during the shutdown of March 2011:

Table 2. Internal inspection of Reboiler A in 03/2011

Bottom Cover	Surface residue
Bottom Mirror	No deposits or internal blockage of the tubes were observed.
Top Mirror	Approximately 30% of the tubes were clogged, especially in the region of the vapor inlet, which has higher temperatures. Magnetic and non-magnetic deposits.

May/2011	**May/2011**

Figure 3. The mirror of Reboiler A before it was cleaned.	**Figure 4.** The mirror of Reboiler A at the vapor inlet before it was cleaned.

Note: It is difficult to carry out internal cleansing by Hydro jetting (pump pressure during cleaning is 20,000 psi). This reboiler has 11 tubes plugged into it, but none of them are plugged in during this shutdown.

Campaign II: Production shutdown on 03/2011

Table 3 below summarizes the features observed during the inspection of the state of the inlet of this piece of equipment during the shutdown in May 2012:

Table 3. Internal inspection of Reboiler A in 05/2012

Bottom Cover	Surface residue
Bottom Mirror	No deposits or internal blockage of the tubes were observed.
Top Mirror	Clogged tubes were not identified, not even in the region of the inlet vapor. There was only a thin layer of deposit that was easy to remove and which was evenly distributed over the upper surface of the mirror.
Spool	A thin deposit on the inner surface of the caisson.

May/2012

Figure 5. The mirror of Reboiler A before it was cleaned.

May/2012

Figure 6. The mirror of Reboiler A before it was cleaned.

Note: Hydro jetting was done quickly, without any obstructions due to deposits having been identified. This equipment has 11 tubes plugged into it, but none of them were during this shutdown.

Analysis of Deposits

The characterization of the deposits found in Reboiler A showed an 86.0% loss on ignition, thus indicating the strong presence of organic compounds in their composition, with the functional groups of carboxylic acid and ester standing out. The inorganic analysis of the sample indicated a high concentration of iron oxides and also sulfur oxides in the form of sulfides on the cover and spool of this reboiler. This is summarized in Table 4 below:

Table 4. Analysis of the Deposits of Reboiler A

Organic Compounds - LOI (%)	86.0
Water of hydration	Smaller
Sulfur compounds	Smaller
Carboxylic Acid and Ester	Greater
Dichloromethane soluble (%)	2.0
Elemental Sulfur	Greater
Dichloromethane insoluble (%)	98.0
Water of hydration	Smaller
Inorganics	Smaller
Carboxylic Acid and Ester	Greater
Composition of Dichloromethane insolubles (%)	
Carbon	56.0
Hydrogen	5.0
Nitrogen	3.0
Sulfur	8.0
Inorganic Compounds (%)	
Iron, Fe2O3 + Fe3O4	82.0
Sulfide, S2-	16.0
Phosphate, P2O5	1.0
Sodium, Na2O	1.0

Evaluation of Performance

The evaluation of performance of Reboiler A was prepared using the MRA tool, which is a GE good practice to predict fouling in critical systems. This statistical tool enables a dependent variable to be determined, in this case the percentage of opening the steam valve, by means of the numerical regression of the other operational variables of the column. This method generates a mathematical equation to predict the percentage of inlet steam when the equipment is clean, given that the equation obtained by regression includes data of this condition.

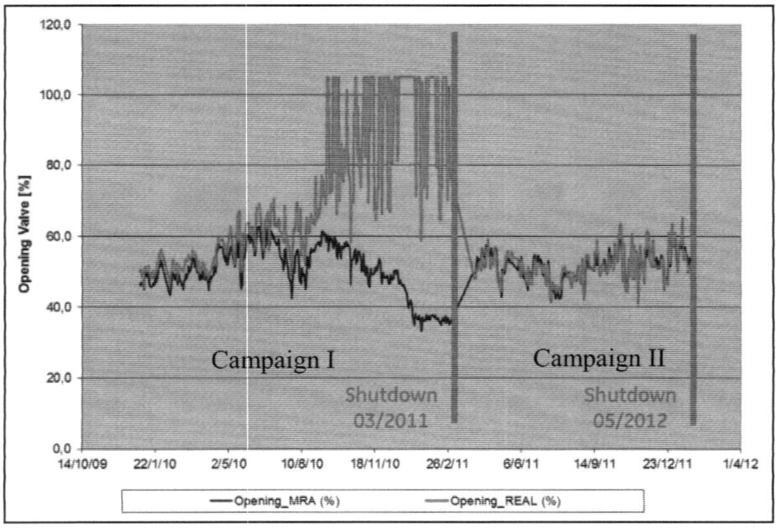

Figure 7. Performance Evaluation of Reboiler A

In Figure 7 the dark blue line represents the percentage of inlet steam when the equipment is clean, as per the equation defined by the MRA. The pink line represents the actual percentage of inlet steam. The sharp upward shift evidenced in the period between November 2010 and March 2011 shows the fouling of the system during Campaign I (prior to the 2011 operational shutdown).

After the 2011 shutdown, it can be seen that the two curves follow the same trend, without a shift, indicating that the unit has the equivalent performance of a clean piece of equipment. The change in the GE treatment strategy, taking the control of the dosages of the product Petroflo 20Y25 into account, was seen to have been effective in increasing the campaign time of this equipment.

Conclusion

The change of strategy in applying dispersant and controlling the dosages after the March 2011 shutdown, presented positive results, given the states of the inlet seen during the inspection of reboilers A and B in May 2012..

The low degree of fouling of these reboilers evidenced in the 2012 shutdown led to the production process being considered reliable, as the A-100 reboilers are no longer the pieces of equipment responsible for the loading limit of the unit. The downtime during the cleaning of the mechanical parts of the reboilers was about 10 times less than that observed in the previous shutdown, thus reducing Braskem´s maintenance costs.

In addition, the MRA statistical tool was able to produce a reliable representation of the degree of fouling of the system. This tool can be effective for scheduling shutdowns more than 03 months in advance.

References

1. Setal Lummus, “Manual de Operação da Unidade de Processamento de Nafta Bruta (Área 100)”, Santo André, Brazil.
2. Ponce de Leon, Carolina; Weissheimer, Flavio and Camparin, Rafael H. (2011), “Proposta Técnica de Fornecimento de Especialidades Químicas, Equipamentos e Serviços para Tratamento Químico para a Unidade de Processamento de Nafta Bruta (Área 100)”, Santo André, Brazil.
3. Kleinbaum, David G., Kupper, Lawrence L., Muller, Keith E. (2008) “Applied Regression Analysis and Other Multivariable Methods”, Belmont, USA.

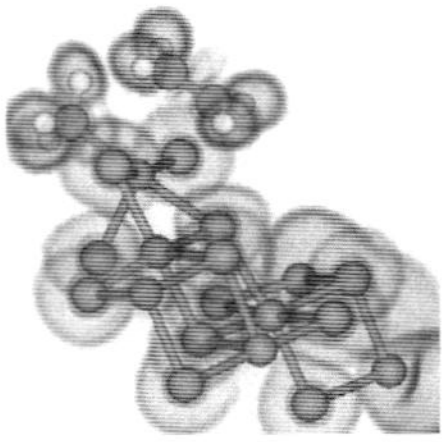

13 AIChE
Spring Meeting, San Antonio, TX

SmartProcess® Application Strategy Achieves Faster and More Efficient Control of an Ethylene Complex

Patrick Truesdale
Principal Consultant
Industry Solution Group
Emerson Process Management

April 29, 2013
San Antonio, TX

EMERSON
Process Management

Speaker Introduction

- Principal Consultant
  Emerson Process Systems & Solutions
- BS Electrical Eng, PE
- 40+ Years experience with Refinery & Petrochem Process Control
- Joined Emerson in 2007
- Previous employers: duPont, Bonner & Moore, SETPOINT, AspenTech, Comcept Technology

Patrick
Truesdale

E-mail: patrick.truesdale@emerson.com

13 AIChE
Spring Meeting, San Antonio, TX

EMERSON
Process Management

Agenda

- Performance Package Review
- SmartProcess Ethylene Overview
  - Furnaces
  - Distillation
- Benefits

SmartProcess Ethylene Packages

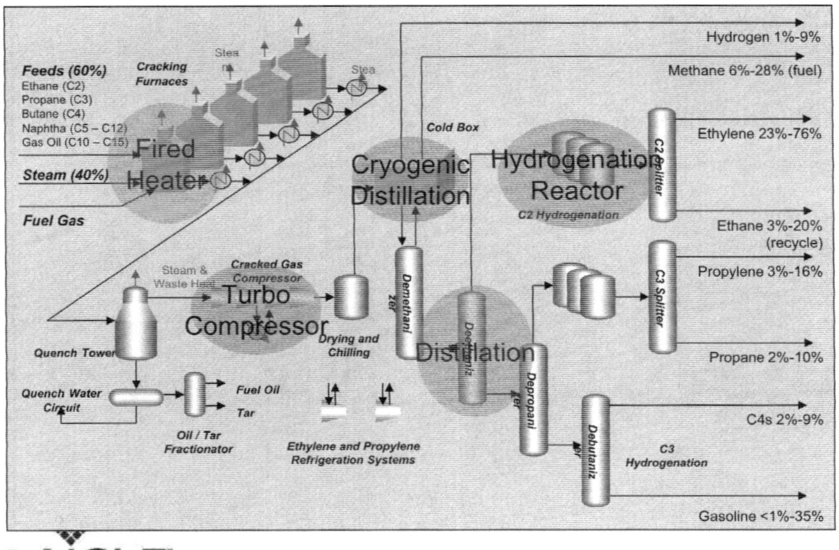

SmartProcess Scope

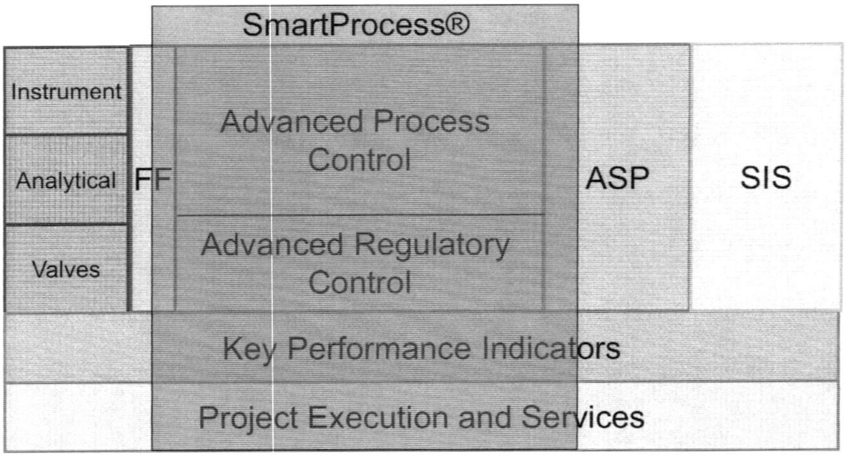

13 AIChE Spring Meeting, San Antonio, TX

EMERSON. Process Management

Basic Ethylene Furnace Variables

Key Operating Objectives:

- **Constant severity**
- **Maintain HC/Stm ratios**
- **Min excess air**
- **Maintain within constraints**
- **Max charge (if needed)**

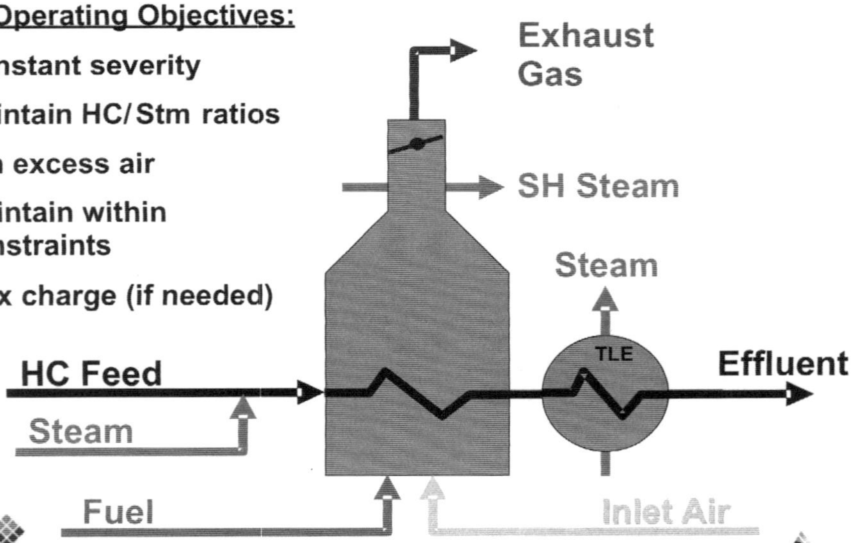

13 AIChE Spring Meeting, San Antonio, TX

EMERSON. Process Management

SmartProcess Furnace Module

- Design allows for different furnace configurations
  - Process Zones
  - Firing Zones
  - Multiple TLE analyzers
- Advanced Regulatory
  - Charge Control
  - Steam Control
  - Severity Calculation
  - Combustion Control
  - KPI's
- Advanced
  - Pass Balancing
  - Heat Balancing
  - Severity Control
  - O2 Control

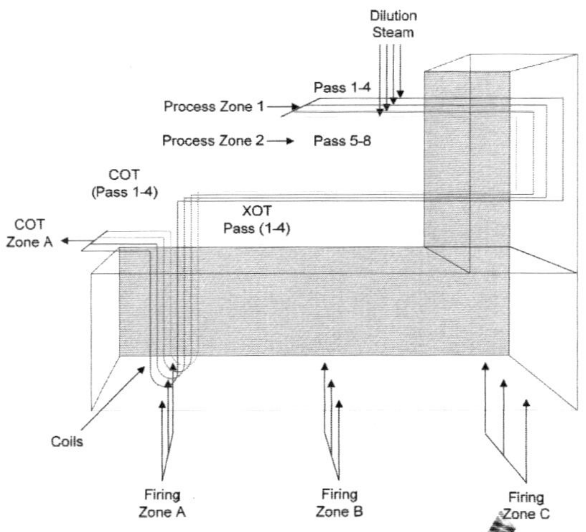

13 AIChE
Spring Meeting, San Antonio, TX

EMERSON. Process Management

Standard Furnace Module Design

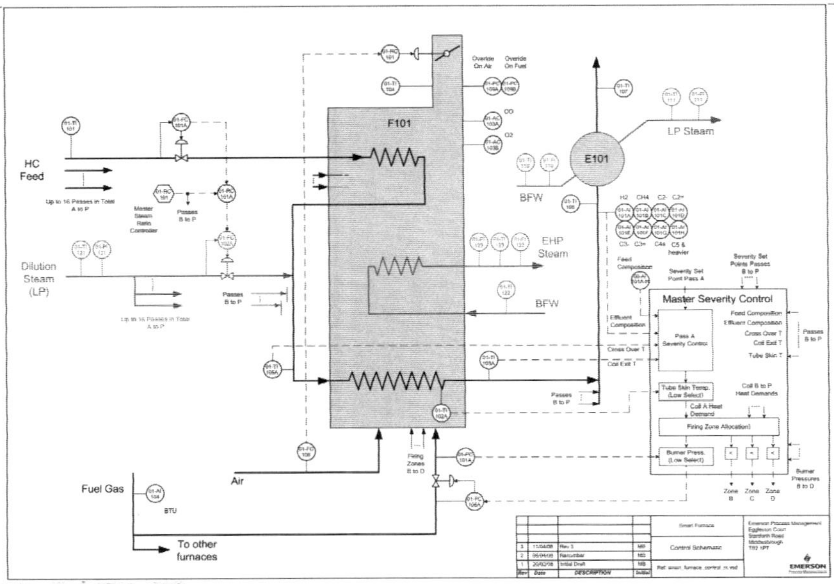

13 AIChE
Spring Meeting, San Antonio, TX

EMERSON. Process Management

Charge Rate Control

Functions:
- Pass Balancing
- Constraints
- Feed Maximization

Functions:
- Ramping
- Limits & Alarms
- Multiple Feed Types
- Furnace Status
- Pass Offsets

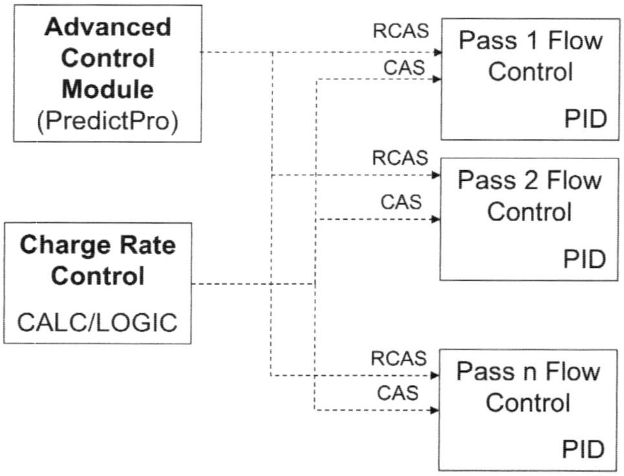

13 AIChE Spring Meeting, San Antonio, TX

EMERSON. Process Management

Steam Ratio Control

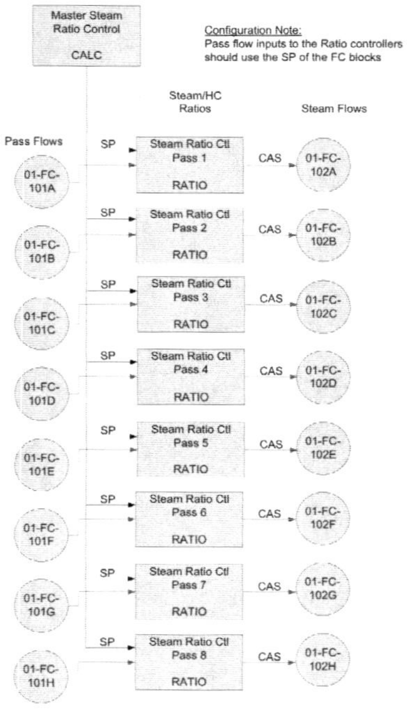

13 AIChE Spring Meeting, San Antonio, TX

EMERSON. Process Management

Severity Control

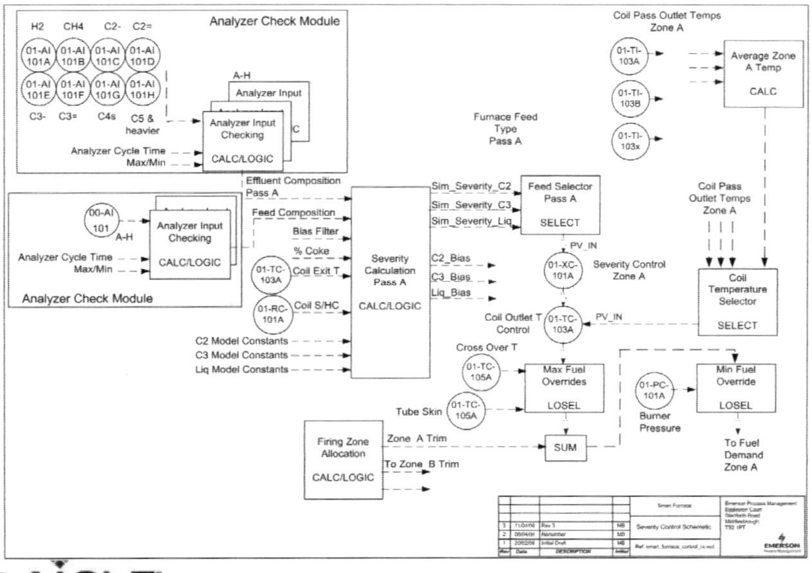

Automated Furnace Sequences

- Standard module templates to be used for Furnace Startup, Shutdown and Decoking sequences.
- Utilizes preconfigured Sequence Function Charts (SFC) running in DeltaV controllers or App Station
- Standard sequence functions can be customized to match site requirements
- Manual operator steps with confirmation can be mixed with automated steps
- Sequence execution control graphics and faceplate templates

Key Performance Calculations

- Furnace efficiency
  - Heat loss method
- Furnace Duty
- TLE Duty, Fouling factor
- Stack gas estimated composition
- Stack gas daily total tons exhaust for CO, CO2, S, N, etc.
- Stack gas NOx emissions (optional – based on CEMS)

13AIChE Spring Meeting, San Antonio, TX

EMERSON. Process Management

Distillation - Classic Multivariable Problem

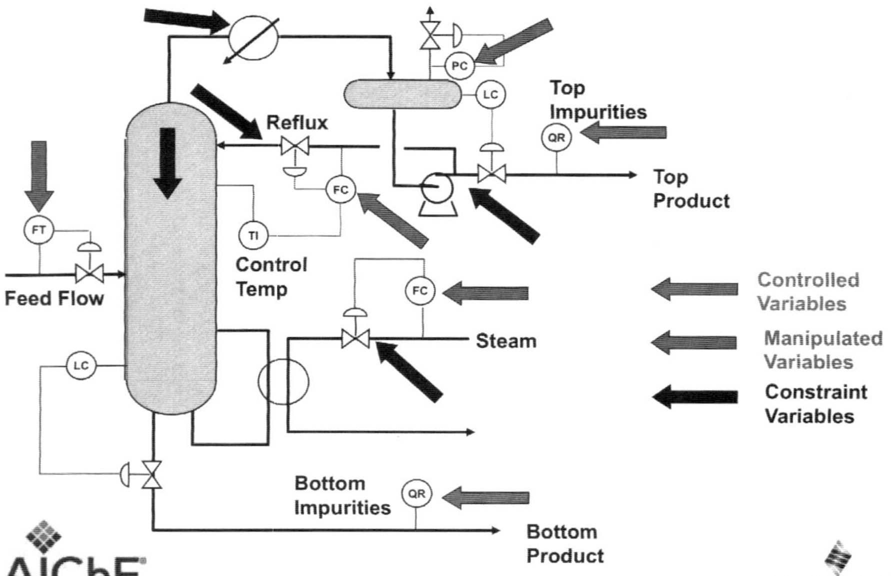

13AIChE Spring Meeting, San Antonio, TX

EMERSON. Process Management

Distillation Control Module

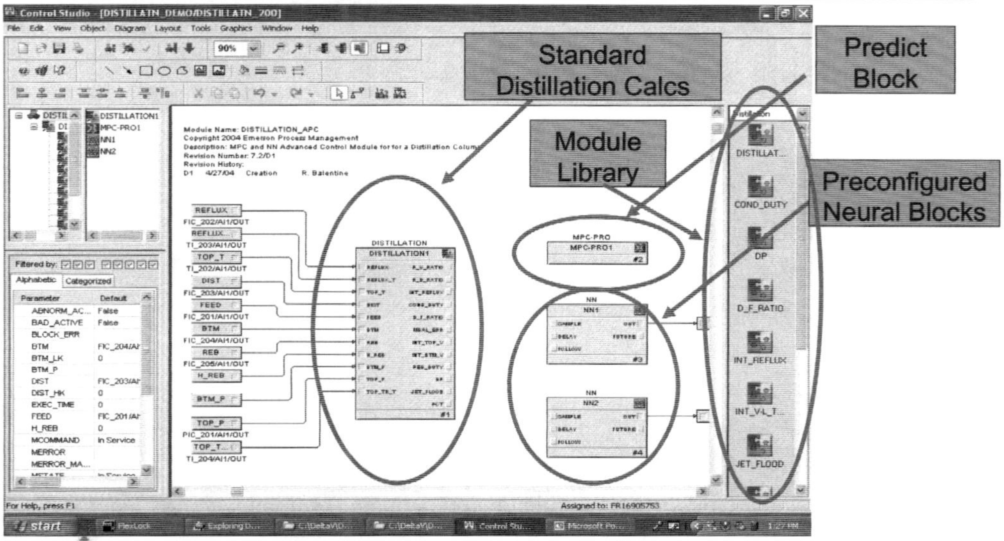

SmartProcess Distillation

- DeltaV PredictPro with up to 4 MV's
- DeltaV Neural for up to 2 product qualities
- Standard set of calculations & constraints
  - Internal reflux
  - Pressure compensated temperatures
  - Jet flooding
  - Reboiler & condenser duty
- Engineering Services to design, implement and commission

SmartProcess Applications

- Licensed industry applications for common processes that are:
  - Pre-engineered, tested, documented
  - Designed by APC Consultants with extensive practical experience
  - Supported & enhanced
- Bundled combination of:
  - DeltaV libraries: composite blocks, modules & templates
  - DeltaV simulation and demonstration example
  - APC tools (e.g. Predict, Neural)
  - Application code (e.g. VB, .Net)
  - Other databases and apps (e.g. SQL*Server, Excel, TankMaster)
  - Sample DeltaV Graphics
  - Documentation
- Templates to be used to configure for specific site requirements
- May be implemented by Emerson or Customer

Benefits

13 AIChE Spring Meeting, San Antonio, TX
© 2013 Emerson Process Management
EMERSON Process Management

Ethylene Furnace High Temperature Imaging Cameras to Monitor Burner Flames ,Tube Temperature and Tube Failure Detection

Thomas M. Canty, P.E.
President
J.M. Canty Inc
Buffalo, NY USA

Abstract

The development of explosion proof, high temperature imaging cameras which operate in ethylene furnaces without cooling air now allows monitoring of ethylene furnaces to improve energy/operating efficiency along with providing important safety benefits. Imaging cameras work off the visual wavelength which provides 2 dimensional view and temperature measurement of the tube temperature which is more accurate than thermocouples and IR probes. In addition visual images of the furnace burners lets you detect flame detachment, smoking burners and flame shape in order to increase burner efficiency and save energy. The safety aspects of the Camera imaging systems is significant as it allows visual detection of pin hole leaks and tube failure in the furnaces.

Introduction

Ethylene furnaces, or any type of cracking unit, rely on thermal management in order to make product efficiently and to operate in a safe manner. Having 'eyes' in the furnace can help to identify problems in their early stages of development. High temperature camera technology is able to survive the environment and give a wide angle view of the tubes from which temperature and tube blockage may be determined. Achievable benefits from a camera system include:

1. Temperature analysis capacity for all areas of the image, not just single point. Multiple zone capability for user customization of analysis.
2. Flame detection and size measurement
3. Detection of tube movement and ability for software to automatically move measurement zones as tubes distorts.

Imaging Fundamentals

Basics of an imaging system include a high resolution sensor, proper illumination, appropriate hardware for the process and a robust software analysis package. For this particular application illumination is created by the process and need only be managed optimally for best results. A high density pixel array yields resolution of the image and measurement as well. The hardware and software are the more challenging aspects of the application and will be discussed in some detail.

Hardware

The design of the camera system is unique in that it's internally exposed lens assemblies are **designed to survive without cooling air**. When installed properly, the combination of system materials and heat transfer coefficients at the camera tip limit heat transfer to the system and allow it to run hot without chance of damage. This feature, when system is installed in a clean, gas burning environment, does not disrupt the temperature or air balances of the furnace because no cooling air is being introduced to the furnace. In addition, the design includes a disposable, quarts shield which protects the process exposed lenses from direct contact with the internal atmosphere. Should any particulate contact the system, the quartz shield is there to protect the optical lenses. Replacement of the protective quartz shield is inexpensive. Replacement of damaged lenses is very expensive and so the Canty design guards against that possibility.

Internally, the system is a two piece assembly consisting of the internal insertion lens and the external camera package. As mentioned earlier in this document, Canty employs a fused glass window as part of the electronic housing package. This has a dual purpose; Firstly, the camera housing assembly is a separate entity which carries WP and EXP ratings. Secondly, the camera housing bolts to the internal lens system and transmits the optical rays through the fused glass window to the CCD. This mounting also creates a hermetic barrier between the process and the electronic housing thereby preventing migration of hot, process gasses outside the pressure boundary of the system.

Figure 1. High Temp Camera Assembly

The camera electronics come with a full array of controls that allow it to be optimized to the software function to achieve accurate, consistent data output.

As previously mentioned, the installation is an important feature of the design. When installed correctly, the camera sits inside the refractory (see Figure 2) where its exposure to the process environment is minimized thus minimizing heat transfer into the internal lens.

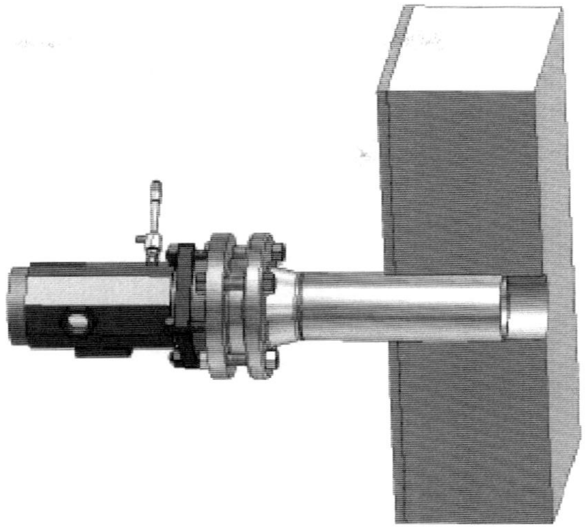

Figure 2. Camera Installation Schematic

Software

With respect to software design, the Canty software is fully flexible in analyzing any portion of the captured image. Multiple analysis zones can be drawn to monitor discreet areas of the view and each zone can run its own particular algorithm depending on need. The benefit here is that multiple tubes can be continually surveyed for changes and trends. The analysis data is also not single point, but is comprised of all the data points in the analysis zone. By choosing proper software filters various information can be obtained from the full array of data such as hot or cold spots. This data can also be archived in total or as snap shots at determined increments of time. The on screen data output can also be updated in the short or near term and can thus be used to judge what is happening in the furnace over time with just a glance at the screen. For instance, using the histogram feature with a 1x daily output a long term trend can be presented on screen in real time.

The basic theory of temperature measurement is based on the radiative emissions of a body at temperature – as the body temperature increases, the radiation curve shifts towards the shorter wavelengths and that is why as a body heats up the eye cannot detect it until it reaches 1000F which is the approximate visible point where emitted radiation is dark red in color. As the temperature increases, the color graduates to bright red, then orange then toward white. A pyrometer functions by measuring a narrow band of wavelength intensity and comparing it to the intensity of a black body to determine the temperature. From the graph below it can be seen that if a single wavelength is monitored the intensity of that measurement changes as the body temperature changes (curve shifts) which then corresponds to a different temperature on the calibration scale.

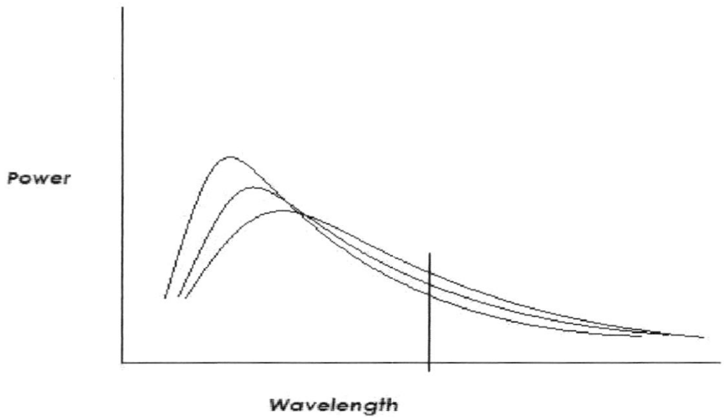

Figure 3. Power Curve at Varying Temperatures

The same is true of a camera system measuring the increasing intensity of an object as it radiates from approx 1000 F (dim, dark red color) on up the temperature scale to 3000+ F for instance which would be a bright white. From the graph above one can see that as color goes from red toward white the peak intensity is always increasing and it is the measure of this intensity which indicates the object temperature. Using an RGB camera the measurement can also be made by monitoring the individual color components of the signal impinging on the chip. For industrial uses, the measurement of intensity usually suffices.

Standard instrumentation used in industry include thermocouple type devices (contact) and thermal imaging and pyrometer systems (non contact). In severe temperatures and atmospheres, the contact devices deteriorate quickly and provide gradually less reliable information leaving the operator running blind. They also create a discontinuity on the surface they are welded to which, in the case of a furnace tube, can lead to premature tube failure.

Non contact systems have difficulties with calibration drift and emissivity adjustments to account for the fact that many items being measured are not black body emitters. In addition, these technologies are single point measurement systems and are best used as hand held devices which can be difficult to use and provide subjective measurement results based on operator preferences for targeting.

Vision technology offers several benefits these other technologies do not. The visual aspect of the system allows the operator to view the process and confirm the readings received. It also allows flexibility to set up measurement zones on screen at any time to assess any or all areas of the process in view should a change be required. Since vision systems are generally fixed installations, an added benefit is that the information is taken from the same process points and features within a given measurement zone at all times which means there is no error in sighting of the instrument.

The vision system is also unique in that it does not suffer calibration drift as almost all instruments do. Once the calibration is set the readings repeat themselves within a couple degrees F. The calibration is based on thermocouple or pyrometer readings initially, however after that calibration is finished it does not have to be redone in the future as long as the system remains in its location.

Process history and performance can also be best documented by a vision system. First off, the camera can analyze any area of process that is within the systems view. This affords the operator the capability of viewing many points of interest instead of just single point temperature readings. As an example, if the system is monitoring a tube bank it can assess the temperature of the tubes for the full length of view. If there is a local flow problem in a tube it will become overheated in one section. The camera can locate that section and indicate its temperature rise over the rest of the tube or adjacent tubes. This information allows the tube location to be identified as well as the position in the tube that has encountered the upset condition. Over time this historical data, recorded in video or still image format by the control software, may be significant in improving the process. In addition to the ability to monitor the full area of the view, the full area of view can also be segmented into multiple zones of measurement, i.e. each tube having its own measurement zone for automated process control capabilities. This capability allows the tube to be operated at optimum efficiency since the addition of excess heat can be avoided thereby reducing energy draw to operate the furnace.

Thermal Analysis

Tube Temperature

The following figures illustrate the temperature analysis capability for tubes in a live test situation. Measurement zones are placed over the tubes (figure 4) as the furnace is heated and cooled. The temperature plots are shown in figures 5 and 6. The results are as expected and uniform from tube to tube, although a small number are included in this demonstration.

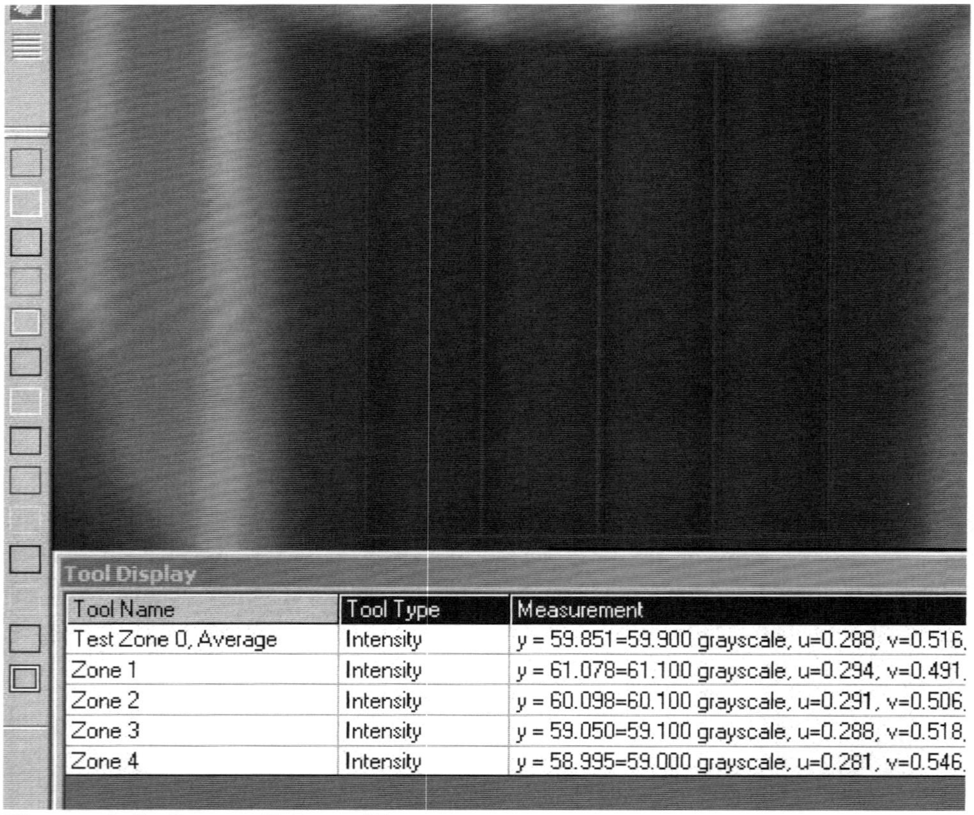

Figure 4. Software Screen Shot

As thermal loads are fluctuated up and down the following graphs track those changes. The cooling cycles are left to natural convection and thus are not linear in nature.

For the following curves, the vertical scale is gray scale intensity which is measured 0 – 255 (pitch black to bright white). The hotter the tube, the higher, or more intense, the gray scale reading. This value in process would be calibrated to the temperature. The horizontal scale is the image number, or time.

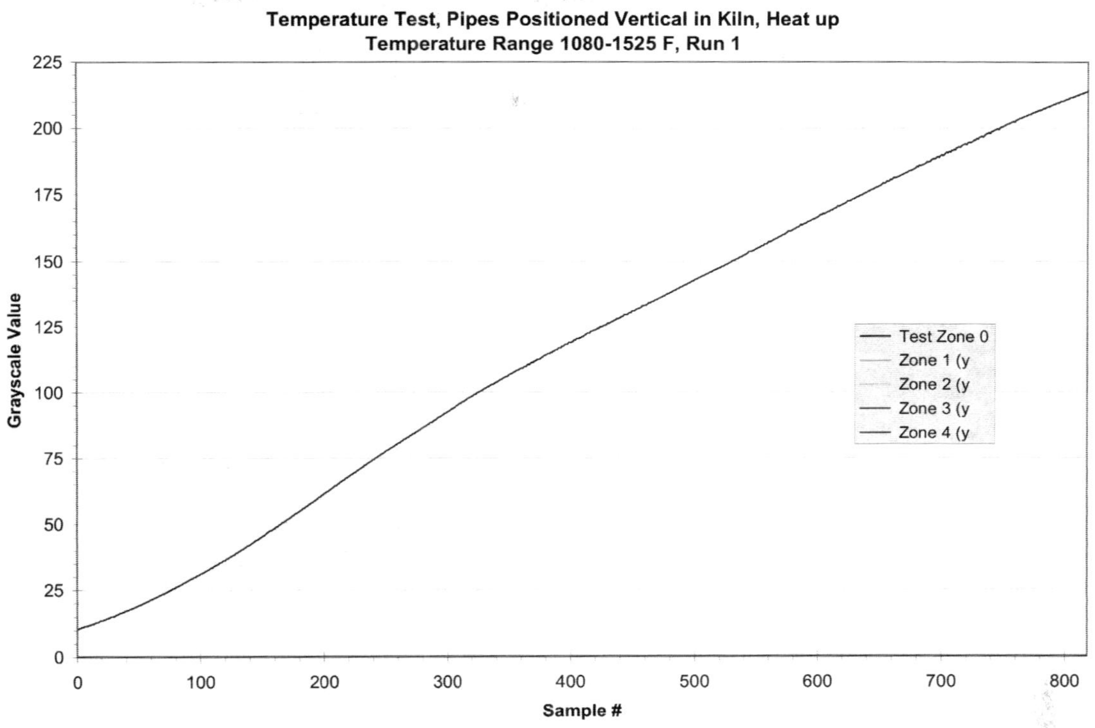

Figure 5. Heating Cycle

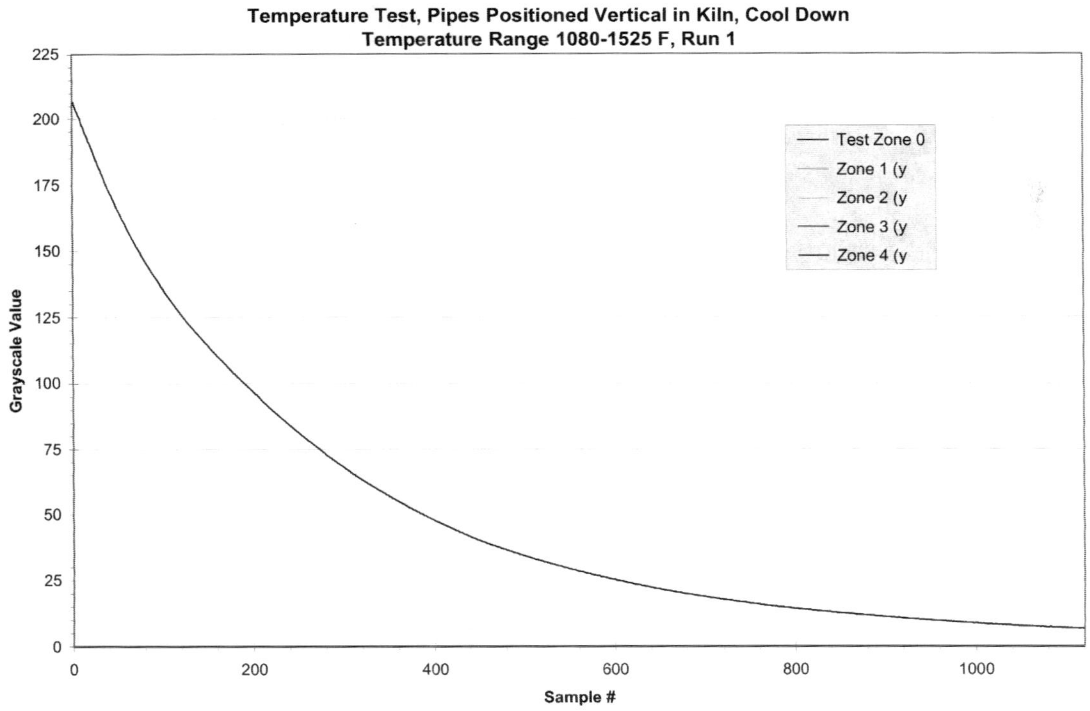

Figure 6. Cooling Cycle

Flame Measurement
Flame temperature can be monitored in the same way as described previously, however the flame size, or presence, can also be monitored. Figure 7 shows a live flame and Figure 8 shows a digital capture. Computing the flame area, or shape, can be done to track its changes over time.

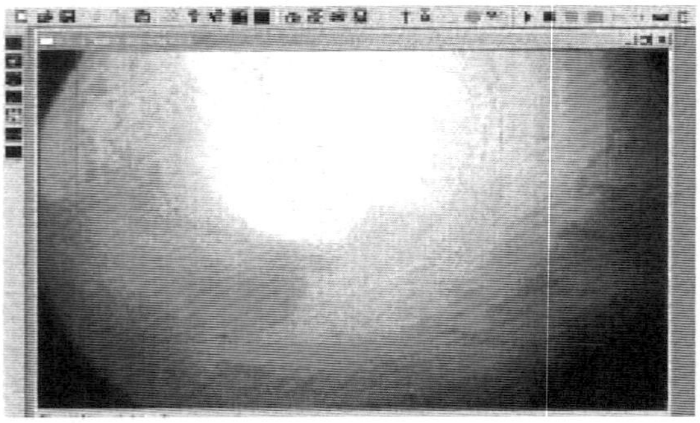

Figure 7. Kiln Flame (Live)

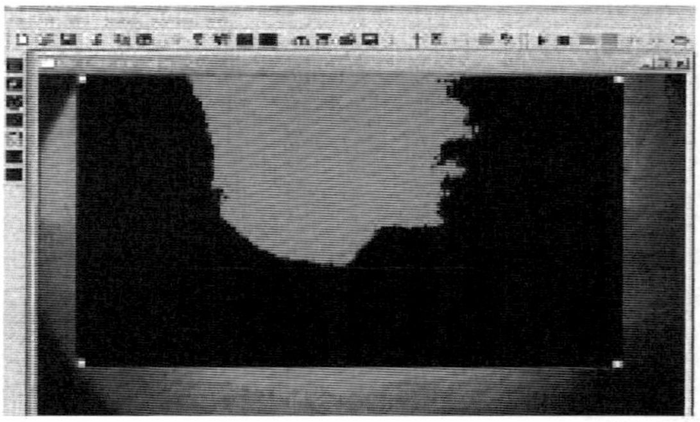

Figure 8. Kiln Flame (Digital)

Conclusion

Vision Technology presents an opportunity to increase effectiveness of furnace monitoring through hardware and software capabilities. The thermal design at the core of the hardware enables the system to provide the operator a view into the furnace and feed back on its thermal performance over time. This data can assist in detecting building upset conditions, but also provide a record for study to assist in improving the operational method of the furnace.

(45v) Straighten up! Non-Linear Compensation Using Feed Forward Control

Barbara Hamilton, Process Systems and Solutions, Emerson Process Management, Hampden , ME

Non Linear processes occur throughout the petrochemical plant. While the PID loop is surprisingly robust for most situations, there are cases where compensation of these non-linearity's is critical for stable control. Similarly, disturbances can cause major disruptions in control stability.

A method of feed forward control, using data based models, can provide robust control throughout the process range. Historical data are used to build models to drive the output toward the solution with PID control providing the final drive to zero error.

The model is also used to develop dynamic tuning. The result is a simple to configure, robust solution to manage non-linear loop and/or disturbance variations.

(45w) Adsorption of CO2 by Alkaline Activated Ca-Al2O3-SiO2 Material

Naim Faqir, King Fahd University of Petroleum & Minerals, Dhahran, Saudi Arabia and Reyad Shawabkeh, Chemical Engineering, King Fahd University of Petroleum & Minerals, Dhahran, Saudi Arabia

Natural gas contains a slight level of carbon dioxide which leads to several industrial and domestic heating problems. In addition, it is one of main greenhouse gases that contribute for global warming, where its pollution is a wide-spread problem arising from a wide range of industries, leading to complex contaminant in atmosphere. Given the increasing pressures worldwide to minimize the emission of greenhouse gases into atmosphere, CO_2 must be remediated efficiently and cost effectively.

In this short abstract a chemically treated limestone and alumino-silicate material were utilized for removal of CO_2 from natural gas using a combined adsorption-chemical reaction processes. This functionalized Ca-Al_2O_3-SiO_2 material was characterized according to its crystalline phase, surface area, pore size distribution and surface morphology. Then it was tested against different concentrations of CO_2 in CO_2-N_2 gas streams.

Pressure Gauges: Hidden Danger or Visible Safety

by Jeff Placek, WIKA Instrument, LP, Lawrenceville, GA

Introduction

In recent years, refining plants, petrochemical plants and chemical plants have been required to increase capacity, but have been provided with very few upgrades to existing equipment to accommodate the increased capacity requirements. This has caused plants to have increased stress and strain on their systems and operating units. For example, higher pulsating pressures and heavy mechanically-induced vibrations are translating to all areas of the processing equipment, especially pressure instrumentation.

Pressure gauges are one of the most important, yet most overlooked or ignored keys to ensuring the continued safety, productivity and profitability of process plants. Process pressure gauges are used to read pressure on transfer lines, discharge pumps, and other places where it is important to be able to read pressures.

Problems Associated with Misapplied, Faulty or Ignored Gauges

Incorrect, misapplied, damaged, faulty or ignored gauges can lead to a number of issues, including safety problems, downtime, and costly regulatory agency fines and other penalties.

Process Safety:

One group of problems relates to process safety issues, such as loss of containment that leads to fugitive emissions, fires and explosions. Bourdon tubes in malfunctioning gauges can rupture, causing the release of media, such as gasoline, oil or gas. Because this media is often volatile and corrosive, it can cause fires, explosions, or other serious safety issues, such as a toxic gas releases or acids spraying onto workers.

Media releases are common results of misapplied, faulty or ignored gauges. Fires and explosions are less common but they do occur. When reviewing plant instrumentation, findings have revealed instances where the wrong gauges have been installed in locations on pumps. In one instance, a gauge vibrated so severely when the pump came out of alignment that it caused the Bourdon tube in the gauge to crack. The media coming out of the tube leaked onto the pump, which was excessively hot because of age and misalignment, causing a fire.

Expensive Downtime:

A second group of problems relates to expensive downtime, which can result from the breakdown of process safety just mentioned, or simply from having to shut down a process in order to replace incorrect, misapplied, broken or otherwise damaged gauges.

1 - Safety-Related Downtime:

As a result of the problems identified above (emissions, fires, explosions, etc.), there can be any number of causes of subsequent downtime, including fire suppression activities, released media containment, line shutdown, personnel evacuation, and then eventually having to rebuild or repair the lines, replace equipment, etc. In addition, if there is a media release with no injuries or fatalities, the plant is required to notify the EPA. However, if there is a media release that does cause injuries or deaths, the plant is required to notify OSHA. These can lead to very long process shutdowns, anywhere from a day to a month or more. (More on OSHA and the EPA below.)

2 - Non-Safety-Related Downtime:

When a line fails, causing a gauge to fail, even if there is no release or safety incident, downtime can still occur as a result of the process needing to be shut down in order to replace the gauge. Gauge failure rarely causes other parts of the process to fail, although gauge failure is usually an alert that there is a problem. In most cases, a plant worker will simply replace the pressure gauge and move on. When another failure occurs, the worker will do the same, until the root cause of the failure, such as a misaligned pump, causes the plant to shut down the line, resulting in more costly, unplanned downtime and a major maintenance time.

Regulatory Agency Fines and Penalties:

A third group of problems relates to the potential for significant fines and other penalties from federal regulatory agencies (primarily OSHA and the EPA), as well as from state and local jurisdictions, that result from compromised process safety.

1 - OSHA:

OSHA addresses many process-related safety issues in 1910 Subpart H, "Hazardous Materials." More specific is 1910.119, "Process safety management of highly hazardous chemicals." In its Process Safety Management section, OSHA notes that, "Unexpected releases of toxic, reactive, or flammable liquids and gases in processes involving highly hazardous chemicals have been reported for many years in various industries that use chemicals with such properties. Regardless of the industry that uses these hazardous chemicals, there is a potential for an accidental release at times they are not properly controlled, creating the possibility of disaster."

OSHA goes on to note that: "To help ensure safe and healthful workforces, OSHA has issued the Process Safety Management of Highly Hazardous Chemicals standard (29 CFR 1910.119), which contains requirements for the management of hazards associated with processes using highly hazardous chemicals."

OSHA's standard emphasizes the management of hazards associated with highly hazardous chemicals and establishes a comprehensive management program that integrates technologies, procedures, and management practices.

While OSHA does not monitor gauges in specifically, they will enforce guidelines for worker safety. OSHA investigates injuries and deaths, and may levy fines or other penalties.

2 - EPA:

According to Greg Neneman, director of business development solutions for Valin Corporation (San Jose, Calif.), fugitive emission regulations from the Federal level have tightened up over the past several years. In many states, such as California, Oregon and Washington, the regulations are even tighter.

"One challenge companies face is ensuring compliance with emission regulations when they upgrade or expand their processes," points out Neneman. "They are not allowed to produce additional emissions beyond the original limit set by regulations."

From the standpoint of federal EPA regulations, each company that produces or emits volatile organic compounds (VOCs) is required to come up with a leak detection and repair program. "In some areas, county and local requirements are even tighter than federal requirements," continues Neneman.

In certain areas, companies that produce or emit VOCs are required by regulation to do audits on valves, flanges, gauges, and all other process connections on a quarterly basis. "And in some jurisdictions, the regulatory agencies can give plants 24 hours notice before they descend on that plant to check all of their records and conduct other investigations," adds Neneman.

In May 2012, the U.S. Environmental Protection Agency (EPA) levied a $1 million civil penalty on a plastics company for alleged violations of the Clean Air Act at two of its chemical manufacturing plants. In addition to paying the penalty, the company agreed to implement a comprehensive program to reduce fugitive emissions of hazardous air pollutants (HAPs) from leaking equipment such as valves and pumps. The settlement also required the company to implement enhanced work practices, including more frequent leak monitoring, better repair practices, and new efforts designed to prevent leaks. After being fined by the EPA, the company realized that instrumentation was one of the areas where they having fugitive emissions. Until that time, workers had been replacing gauges left and right, because they weren't aware of the problems that could occur. The gauges weren't a big expense to them, so they considered the replacement activity to be more of a nuisance than anything. It's when they had the rupture that they realized there was a real problem.

An audit of the plant, which involved reviewing at all of the instrumentation, found that in many applications, they were using the wrong instrumentation. To correct the issues, the right instrumentation was specified for them, and the correct gauges were ordered. The plant now has its instrumentation under control.

While conforming to EPA regulations has always been a challenge for process facilities, it is going to become even more challenging in the future as a result of EPA's recent introduction of new requirements related to fugitive emissions and leak detection and repair (LDAR) strategies. On April 17, 2012, the EPA Administrator signed a suite of final air quality regulations that affect both upstream and midstream oil and gas sectors - the production/processing and transmission/storage sectors of the oil and natural gas industry.

For the first time, EPA will regulate air emissions from natural gas wells that are hydraulically fractured, as well as other emission sources associated with exploration, production, processing, and transportation of oil and natural gas.

Promulgation of the final rules will result in new control equipment requirements, as well as increased requirements for monitoring, recordkeeping, and reporting for owners, operators, contractors and vendors.

The final rules require the use of emission controls, as well as work practices, through two different air regulatory programs. One is the New Source Performance Standards (NSPS). The other is the National Emissions Standards for Hazardous Air Pollutants (NESHAP).

EPA finalized a new NSPS at 40 CFR Part 60, Subpart OOOO, which regulates emissions of volatile organic compounds (VOC) and sulfur dioxide (SO2) from a variety of oil and gas exploration, production, processing, and transportation facilities, including natural gas wells, certain compressors and pneumatic controllers, storage vessels, and sweetening units at onshore natural gas processing plants. Subpart OOOO also includes startup, shutdown and malfunction provisions.

The EPA has also made it clear that oil and gas production and processing sources will be the focus of its National Enforcement Initiative (NEI) for fiscal years 2012 and 2013. With the new NSPS and revised NESHAPs in place, EPA now has the tools it needs to enforce in a comprehensive manner.

As a result of these new regulations, plants should look at everything more closely, including gauges.

How Problems Can Occur With Gauges

Technical problems can occur with process pressure gauges in a number of ways, which can be conveniently categorized into seven broad categories with specialties or subsets. The most common reason for gauge problems is that workers often install gauges without knowing how they should be installed or accessorized. As a result, the gauges can fail because they were never designed to function in those specific situations. They can also fail because they were installed incorrectly, or because they were not properly maintained.

One way to eliminate or minimize these problems is to consider specific pressure gauge accessories. These can prove to be the extra step needed to ensure a successful, proactive preventive maintenance program, and for extending the life of the gauges. Accessories can provide additional protection for the gauges during their lifetimes.

1 - Mechanical Vibration.

This is oscillating, reciprocating or other periodic motion of a rigid or elastic body or medium forced from a position or state of equilibrium. Typically, vibration is the main cause of pressure gauge failure in process plants. Pump applications produce most of this vibration.

Gauges that aren't designed to withstand vibration may be installed in applications where vibration occurs, causing the gauges to come out of accuracy. Even more serious, a gauge can vibrate so much that the Bourdon tube inside it can crack, releasing media."

One way to prevent vibration-related problems is to install liquid-filled gauges, which help to dampen the vibration and also protect the internal components. The liquid case filling also lubricates the internals and prevents corrosion.

2 - Pulsation.

When media rapidly cycles through the pressure system of the gauge, pulsation occurs. If the pointer pulsation increments are greater than 5 percent of the full scale value, you must intervene to prevent damage to the gauge.

A restrictor or throttle screw can solve most pulsation problems. It will slow down the media by reducing the size of the orifice and averaging out pressure fluctuations. For extreme pulsation, such as spiking and surging, a snubber works best. It allows more control over the amount of snubbing and the flow of media into the pressure system.

3 - Temperature:

If a gauge is exposed to a media with very high temperature, the metal alloy in the gauge may not be able to handle the temperature, causing the Bourdon tube to fail and release media.

Long pipes that separate the pressure gauge from the process, or small cooling elements with radiator fins, protect the pressure instrument from high process temperatures by allowing ambient air to flow over the surface area. In environments with steam, using a siphon with the pressure gauge cools the steam down and drives it back to its liquid state before it gets to the gauge. With very high-temperature media, a diaphragm seal with a cooling element will isolate and prevent the hot media from reaching the gauge.

4 - Spikes or Overpressure.

In some cases, process media may be under extreme pressure. Pumps also can cause spikes and over-pressure, both of which shorten gauge life. Gauges can fail as a result of these pressure spikes. When workers turn pumps on or off, or open or close valves, a surge of media or oil goes through the pipe and hits the pressure gauge, which causes a huge spike. For example, if a line is pumping 100 psi of oil, someone may install a 100 psi gauge, not realizing that this leaves no allowance for any overpressure, surging, or spiking. When pressure does increase, it causes overpressure on the gauge, sometimes to the point of where the Bourdon tube ruptures and leaks media.

One way to prevent this problem is to use a different type of gauge, such as 200 psi gauge for a 100 psi line. Another option is to accessorize the gauge with an over-pressure protector, which would shut off the pressure if there were a surge or spike. Overpressure protectors are designed to avert spikes and the resulting overpressure damage. These protectors provide protection for pressure gauges against the effects of pressures exceeding their maximum pressure ratings.

5 - Corrosion.

Corrosion problems can occur with gauges if the process media is not compatible with the pressure gauges' wetted parts material. Media can sometimes be so corrosive that it corrodes through the Bourdon tube and leaks onto the ground.

When the media is not compatible with the wetted parts material of the pressure gauge, a diaphragm seal is required to isolate the gauge from the media. The benefit of a diaphragm seal is that it can be made of special alloys that are designed to withstand corrosion. It is essential to select a diaphragm material that is compatible to the process media.

6 - Clogging.

In some processes, the media may eventually crystallize or may otherwise harden. If a standard gauge is used in such an application, it will eventually get clogged and will not be able to read the pressure accurately. For example, if a gauge gets clogged and ends up reading zero pressure, when in fact it is under tremendous pressure, there will be an unexpected and dangerous release of media if and when the piping system is opened. Diaphragm seals protect the pressure gauge from solids, particles and viscous media by creating a barrier between the process and the pressure inlet of the gauge. There is also an option to equip the diaphragm seal with flushing ports to constantly flush material out from the diaphragm surface.

7 - Mishandling and Misapplication.

Gauges can also fail prematurely if they are misapplied. For example, a gauge may have been ideal for the application when it was installed. However, if there are equipment modifications or more components subsequently added, this can change process parameters, resulting in the gauge not longer being right for that application. Misapplication can lead to serious problems associated with failure and subsequent media release.

It is also possible to accidentally damage gauges during installation. Gauges should be installed using the proper sized wrench on the flats. Gauges should never be installed by turning the case. After installation, gauges can also fail if they are mishandled or used improperly.

Why Problems Can Occur With Gauges

In a typical process plant, about 25 percent of pressure gauges have failed or are about to fail.[1]
A lack of management focus and worker education and experience are key reasons plants experience issues with gauges.

Management Focus.

Gauges are often viewed as inexpensive commodities, with management failing to realize that they are very critical parts of the processes. Often, plants will replace failed gauges with the same replacement gauges, which will also fail, only to have to be replaced again. Eventually, their luck may run out and a serious failure may occur because the incorrect gauge has been installed.

Why do gauges tend to receive such little attention from management? Management tends to focus on the expensive equipment - $100,000 or million dollars items. They often don't understand the importance of focusing attention on $100 gauges. This is not the fault of management. The problem is that no one has explained to them just how important gauges are.

As noted earlier, though, ignoring instrumentation can cause serious safety and financial problems. Plant process gauges can be thought of in the same way as vehicle gauges. If the temperature gauge on an automobile is defective, it is possible to destroy the engine as a result of overheating that is not registered by the defective temperature gauge. Similarly, if the oil pressure gauge on an automobile is defective, it is again possible to destroy the engine as a result of seized components resulting from loss of oil that is not indicated by the defective oil pressure gauge.

Worker Education and Experience.

A lack of worker education and experience also contributes to gauge issues. In many instances, when maintenance technicians look at gauges, they don't pay much attention to them unless they leaking or look as if they are going to explode. They don't realize all of the information that that gauge can provide. In addition, technicians may not realize that a gauge has failed or is about to fail. Again, as long as it is not leaking media, they will often just keep walking. This is rarely the fault of the technicians. Many have not been properly trained to understand process pressure gauge performance.

Along the same lines, the industry has been experiencing a "brain drain" for several years, with large numbers of experienced technical workers and engineers retiring and being replaced by younger people who lack the education and experience to understand what gauges are telling them, or, worse, not being replaced at all.

According to research conducted by Cambridge Energy Research Associates in 2007, more than half of all oil-field professionals will reach retirement age during the next decade. The Interstate Oil & Gas Compact Commission found that the average age of production and service company employees is 46 to 49 years, and the average retirement age in the oil and gas industry is 55 years.[2]

Historically, plants had specialists whose jobs focused exclusively on instrumentation. They spec'd the gauges, maintained them, and made sure they were operating correctly. However, as plants got leaner in recent years, they started letting these people go. Plants first shifted gauge responsibility to the pipefitter whose main responsibility was to make sure the pipes were put together correctly and not leaking.

[1]WIKA data aggregated from more than 250 instrument audits.
[2]Peter Peek, Jocelyn Fenard, Patrick Gantes, Christine Theiler. December 2008, "Skills Shortage in the Global Oil and Gas Industry: How to close the gap" Part I. Pages 14-15.

Then, as plants began letting pipefitters go, they shifted the responsibility to the unit operators or technicians, who now install the gauges. These people usually don't have any experience or knowledge of how to install instrumentation. They usually just replace old gauges with similar gauges. The problem is that the gauges are usually the wrong ones to begin with. The result is that, these days, few plants have people who have much if any education and experience with instrumentation.

Taking Stock of Process Pressure Gauges

When it comes to improving the safety, productivity and profitability of process plants through an effective gauge management program, companies can take a couple of steps. First, an understanding of the vital importance of gauges must be established. Second, workers must understand what kind of gauges will be effective.

Focus on Gauges:

World-class process plants maintain their assets proactively. In such plants, pressure gauges receive as much attention as pumps and control valves. Management understands the importance of proactively maintaining their assets, and they realize that one of the best ways to maintain their most expensive assets is to properly source and maintain the gauges that protect these assets. That is, they understand the value of spending $100 for a gauge that can subsequently prevent damage or destruction to a $100,000 or $500,000 piece of equipment and the resulting negative safety, downtime and regulatory implications.

There is an interest in predictive maintenance these days in process plants, which is a good thing. A lot of this, though, involves expensive software. In addition, when something is out of kilter, the system may be set up such that technicians may have to wait for it to get far enough out of the normal range before it will generate an alert on the computer. However, properly-maintained pressure gauges, studied by people who know what to look for, can be very inexpensive but reliable and useful mechanical predictive maintenance tools.

Gauge Procurement Considerations:

Over time, as noted earlier, the status of pressure gauges has unfortunately diminished to that of commodities in many process plants. In order to ensure process plant safety, productivity and profitability, it is essential that plant professionals understand the importance of selecting the right gauges for applications.

It is also important that these be high-quality gauges, not "commodity" gauges. That is, not all process gauges are created equal. The so-called "lowest price" gauges appear attractive at first, but are rarely the "lowest cost" in the long run, because they usually have to be replaced more often, leading to excessive downtime and inconvenience. Worse, they may fail unexpectedly, leading to catastrophic results.

Consider these 10 steps to select the right gauges that will help to ensure safe, productive and profitable plant processes.

1 - Know what is in the pipe. This way, you can select a gauge alloy that will be compatible with the process media.

2 - Next, know the temperature of the media. If it is very hot, for example, it will be unsafe to use a standard pressure gauge. A long pipe, siphon or diaphragm seal assembly may be needed to cool the media sufficiently before it reaches the gauge internals.

3 – Third, know the ambient conditions surrounding the gauge. For example, if it is corrosive around the gauge, you need to make sure that the external components of the gauge such as the case and window can handle the environmental conditions.

4 - Next, know the pressure in the piping that the gauge is meant to measure. A best practice is using a gauge that is designed for two times the normal operating pressure. For example, if normal pressure is 100 psi, use a 200 psi gauge.

5 – Know the conditions that affect gauge wear. These could include mechanical vibration, pulsation, pressure spikes, etc. For example, if vibration exists, a gauge with a liquid-filled case is recommended. If there is pulsation, use a throttle plug or restrictor accessory.

6 - Select gauges that are designed for a particular service. Process type gauges are designed with strong Bourdon tubes and other heavy duty components. If a gauge is misapplied, the internals of the gauge may not be rugged enough to resist failure and subsequent media release in these applications.

7 - The gauge case must be rugged and have the correct safety features. In some case designs, if there is a rupture of the Bourdon tube, media will be released out the front of the gauge, which can spray workers. Select a gauge where media would release out from the back of the gauge.

8 - Accuracy is important. In most process plants, +/-0.5% accuracy is the norm, and +/-1% is acceptable.

9 - Size is also important. The gauge dial needs to be large enough that operators can read it from a distance – four inches or larger.

10 - Finally, the gauge should be mounted properly. A gauge is usually mounted by its stem, but it can also be mounted in a panel or against a flat surface.

Looking Outside for Assistance

As noted earlier, it can be difficult for plant management to devote enough time and attention to an effective process pressure gauge program. Similarly, it can be difficult for technical personnel to get up-to-speed in terms of education and experience related to process pressure gauges.

There has been a growing trend over the last 30 years in virtually all industries for companies to outsource services that are not considered "core competencies." That is, companies have come to realize that it makes sense to focus their attention on their missions and what they do best, and leave secondary and peripheral activities to third parties that specialize in these other activities and services. Becoming experts in pressure gauges is certainly not a core competency for refining plants, petrochemical, and chemical plants, so this is an area that can easily be outsourced to an experienced third-party audit team.

If a plant elects not to gain the experience that it needs on its own, free assistance is available. To ensure that gauges are working properly and are able to protect the valuable assets that they were designed to protect, a five-phase instrument review is recommended. A best practices review identifies gauges with unwanted conditions and recommends corrective actions to ensure optimum performance. The review identifies issues that affect reliability, longevity and safety. Such a review can uncover evidence of pulsation, vibration, pressure spikes, excess process or ambient temperature, installation damage, clogging, chemical incompatibility, and other problems that can lead to serious problems.

Phase 1 - Interview. This involves discussions between the third-party audit team and plant personnel. During the interview, plant personnel learn the purpose of the instrument audit, how it will be conducted, and how they will benefit from the audit. The interview also helps determine the best way to match the audit to the needs of that specific plant.

Phase 2 - Pre-Meeting. This involves determining the units or sections of the plant to audit, as well as outline roles and responsibilities. It also identifies prohibited and exempt areas of the plant. The third-party team reviews problematic instrument installations and discusses safety procedures with plant personnel.

Phase 3 - Instrument Application Audit. This involves having key plant personnel lead the third-party team on an actual plant walk-through. The audit team reviews the target gauge installations, identifies gauges that demonstrate unwanted conditions, as well as those that demonstrate installation best practices.

Phase 4 - Post-Meeting. This involves explaining the instrument audit results to key plant personnel, specifically noting gauge installations that pose immediate safety concerns. Together, the audit team and plant personnel determine which installations to investigate immediately and which to block out. The audit team and plant personnel then set schedules for the follow-up meeting.

Phase 5 - Follow-up Meeting and Presentation. The third-party team presents its general findings and the executive summary resulting from the instrument audit. The team also presents a spreadsheet that lists details about the gauges from the instrument review, color-coded by severity. The team also provides recommendations for improvements based on experience and best practices. Together, the audit team and plant personnel develop a strategy for moving forward.

The plant should also consider leveraging the audit team in other ways. For example, over time a plant's storeroom inventory can get out of hand. A qualified third-party audit team can help plant personnel rationalize, consolidate and manage their inventory of pressure gauges. For example, if a plant stocks spares for 10 or more ranges, the audit team may be able to help reduce that number to five or six ranges.

Essentially, when used properly, gauges play a significant role in plant safety, productivity and profitability. They act like silent alarms, indicating when something is seriously wrong or helping to diagnose deeper issues within systems and processes that can eventually lead to serious safety issues. To ensure safety and reliability at your plant, make certain the proper gauges are in place by reviewing your processes, gauges and your inventory. Everything is at stake.

REFERENCES

1. WIKA data aggregated from more than 250 instrument audits.

2. Peter Peek, Jocelyn Fenard, Patrick Gantes, Christine Theiler. December 2008, "Skills Shortage in the Global Oil and Gas Industry: How to close the gap" Part I. Pages 14-15.

(53a) Sources Of Energy (from Wood to Hydrogen), How Did We Get Here and Where Do We Go Now

Terry Payne, Oak Ridge National Laboratory, Oak Ridge, TN

Energy is simply defined as the ability to do work. Energy provides lights for cities and powers automobiles, trains, jets, machinery in factories and tractors on farms. Energy warms homes, cooks food, and is generally responsible for what is defined as a better quality of life. It is generally believed that a better quality of life should be provided for the next generation and generations to come. Thus, over the past 40 years, the world need for energy has increased geometrically.

However, all sources of energy discovered thus far are finite in quantity and as they are implemented, the supply of the source begins to be depleted. This paper will introduce this session and will trace the utilization of primary sources of energy during the past 40 years, including chemical engineering contributions to development of technologies that have led to practical utilization of these sources and a brief discussion of how chemical engineering is poised to play a significant role in the utilization of energy sources of the future.

(53b) Component Degradation and Its Mitigation In Polymer Electrolyte Fuel Cells

Vijay Ramani, Chemical Engineering, Illinois Institute of Technology, Chicago, IL

Two primary degradation modes that currently limit fuel cell stack lifetime and considerably increase stack cost due to the required over-engineering are: **1)** the corrosion of electrocatalyst support during transient operating modes (e.g. start-up / shut-down) resulting in electrode thinning and loss of electrocatalytic activity, and **2)** the free-radical-induced oxidative degradation of the electrolyte membrane during PEFC operation, resulting in membrane thinning and pin-hole formation. This talk will provide a brief overview of each degradation mode, explaining how they can led to catastrophic stack failure, and will then present the advances made in our laboratory to offer effective and novel solutions to these seemingly intractable problems.

To mitigate support corrosion, a new class of mixed-metal-oxide supports have been synthesized, characterized, and evaluated in our laboratory. Testing both at IIT and at an automotive OEM has confirmed the exceptional corrosion resistance of this class of supports. To mitigate electrolyte degradation, a novel ***in-situ***-fluorescence-spectroscopy-based technique has been developed in our laboratory to monitor, for the first time, the rate of reactive oxygen species (ROS, specifically OH* and OOH*) formation within the electrolyte of an operating PEFC. Remarkably effective regenerative free radical scavengers have been synthesized and evaluated in our laboratory; these materials, when added to the polymer electrolyte, have lowered the macroscopic rate of oxidative polymer electrolyte degradation by close to 2 orders of magnitude.

(53c) Investigating the Role of Ion in Methanol Oxidation by MDH for Bio-Fuel Cell Applications

Purnima Kharidehal, Louisiana Tech University, Ruston, LA

Metalloenzymes have been considered as molecular electrocatalysts due to their extraordinary characteristics. However, enzymatic fuel cells have been reported to have power output and stability limitations; which are restricting the use of this kind of fuel cell to small electronic devices. Methanol dehydrogenase (MDH) is one such enzyme, which oxidizes methanol and other primary alcohols to their corresponding aldehydes and the active site of MDH contains a divalent cation (Ca^{2+}). Ca^{2+} ion holds the PQQ in place, and also acts as a Lewis acid, contributing to the methanol electro-oxidation reaction mechanism by this enzyme. From the proposed mechanisms for methanol oxidation by MDH in the literature, the Hydride Transfer (H-T) mechanism seems, to the best of our knowledge, to be the preferred one under normal conditions. It was also demonstrated that the divalent cation has a central role in the proton abstraction catalysis and, replacing Ca^{2+} with other divalent ion could modify the methanol oxidation pathways. Work reported in the literature shows that binding of substrate and reaction energy barrier for substrate oxidation by dehydrogenase enzymes is influenced by the nature of the ion in the enzyme active site. Thus, understanding the role of the ion in the active site of MDH, as well as the methanol oxidation mechanism may have major impacts on alternative power sources research as they could lead to the development of new bio-inspired synthetic catalysts that could impact the use of methanol as fuel.

In this study, the binding of methanol to the active site models of ion modified MDH is determined and the effect of ion on methanol oxidation is investigated. It has been observed that the binding of methanol and free energy barrier for the rate determining step of the H-T mechanism decreases as the ionic size increases. This shows that replacing the naturally occurring ion (Ca^{2+}) with Mg^{2+}, Sr^{2+} and Ba^{2+} effects the methanol oxidation process and binding of methanol to active site of MDH. Density Functional Theory (DFT) calculations at BLYP/DNP theory level are performed using the $DMOL^3$ module of the Materials Studio software to evaluate binding energies and investigate the reaction pathways. Further, polarization curves corresponding to the electrochemical methanol oxidation in bio fuel cell anodic chambers when MDH enzymes are used as the anode catalysts are obtained using kinetic Monte Carlo approach. Microscopic reaction rates, obtained from free energy barriers evaluated using DFT and transition state theory (TST), will be provided as inputs in a Kinetic Monte Carlo (kMC) program (CARLOS 4.4) to model the oxidation process at macroscopic level. These simulations will give a better understanding of the catalytic methanol oxidation mechanism by MDH, helping evaluate the enzyme catalysis and their dependence on various factors like the nature of the ion in the MDH active site.

(53d) Redox-Flow Battery Based On a Nonionic Molecular Sieve Zeolite Membrane

Ruidong Yang, Chemical and Materials Engineering, University of Cincinnati, Cincinnati, OH, Zhi Xu, Department of Chemical and Materials Engineering, University of Cincinnati, Cincinnati, OH and Junhang Dong, Chemical Engineering, University of Cincinnati, Cincinnati, OH

The conventional redox-flow batteries (RFBs) use polymeric ion exchange membranes (IEMs) which have inherent shortcomings of long-term crossover of reactive ions and material instability in the highly acidic and oxidizing solutions. Here we demonstrate the thin film of siliceous MFI-type zeolite (i.e. silicalite) as a robust electrolyte membrane for the all-vanadium RFB (VRFB). The crystalline silicalite contains uniform channels of an effective diameter of 0.56 nm that allows for the small H_3O^+ ions to diffuse through but denies entry of the hydrated multivalent metal ions by size exclusion (steric) effect. Unlike the polymeric IEMs, the silicalite membrane is nonionic and its proton conductivity relies on the electrical field-driven H_3O^+ transport. The VRFB equipped with the silicalite membrane has shown reduced self-discharge rate as compared to the polymer IEMs and high Coulombic efficiency in charge-discharge cycles.

(53e) Effects of Nanocarbon Buffer Layers On the Anodic Performance of Copper Oxide and Ball-Milled Graphite Composites

Eun-Suok Oh, University of Ulsan, Ulsan, South Korea and Dong-Won Jung, University of Ulsan

Graphite has been used as an anode active material in lithium ion batteries (LIBs) owing to its stable reversibility for a long cycle life, despite of its low capacity (372 mAh g^{-1}). In order to increase capacity in LIB, tremendous research works have recently given to high capacity 3d metal oxides (MOs) such as Co_3O_4, CuO, NiO, and Fe_2O_3. However, these metal oxides undergo a significant volumetric change during charge/discharge processes, leading to materials pulverizations and ultimately rapid electrode degradation. The MOs were anchored on two-dimensional graphene sheets or on one-dimensional carbon nanotubes (CNTs) to alleviate volume expansion and the agglomeration of MO particles. However, CNT and graphene supports are easily aggregated in electrode manufacturing processes. They also have high irreversible capacities due to the formation of solid electrolyte interface (SEI) film on their large surfaces. Unfortunately, detailed investigations of nanocarbons (graphene or CNT) as buffer layers in anodes have not been performed.

In this presentation, nanocarbon papers will be thoroughly compared as buffers for the volume changes of nano-sized CuO/ball-milled graphite (BMG). Morphological and electrochemical characterization techniques will be used for comparison. In addition, the van der Waals forces between CNT layers or graphene layers will be theoretically calculated and their effect on the performance of the buffer layers will be discussed.

() Contributions of Current Density and Voltage Efficiency to Estimated Capital Costs of an All Vanadium Redox-Flow Battery

Mark Moore, Department Chemical and Biomolecular Engineering, University of Tennessee, Knoxville, Knoxville, TN, Che-nan Sun, University of Tennessee, Knoxville, Knoxville, TN, Thomas Zawodzinski, Department of Chemical and Biomolecular Engineering, University of Tennessee, Knoxville, TN, Robert Counce, Chemical and Biomolecular Engineering, University of Tennessee, Knoxville, TN and Jack Watson, Department of Chemical and Biological Engineering, University of Tennessee, Knoxville, TN

Utility scale flow batteries have the potential to provide a number of important benefits including peak shaving, load leveling, electrical grid stabilization and energy storage from wind or solar power. Research into the all-vanadium redox-flow battery (VRB) seeks to take advantage of the multiple oxidation states of vanadium ions and the diminished risk of cross contamination between the anolyte and catholyte solutions. This presentation focuses on utilizing the latest data from research activities at the University of Tennessee in determining the controlling parameters of capital cost estimation. The purpose of this study is to utilize recent data to find a simple and effective way of predicting the manufacturing cost trends. The single-cell data cover a range of values of current density, and associated voltage efficiency. The experimental data from these tests appear to be the highest current densities yet published. The manufacturing costs of the stacks are inversely proportional to the product of current density and the voltage efficiency, while the cost of the vanadium electrolyte is inversely proportional to the voltage efficiency.

Acknowledgement

We gratefully acknowledge support of the Electric Power Research Institute and the U.S. National Science Foundation (NSF) funded Tennessee Solar Conversion and Storage using Outreach, Research and Education (TN-SCORE) Program, NSF EPS-1004083, under Thrust 2.

Overview of P.E. Licensing Process in United States

William R. Parrish, P.E. and Cory D. Jensen, P.E.

Professional engineering licensure was established in the United States to protect the health, safety and welfare of the public from unscrupulous or incompetent decisions of engineers. Licensure provides a proven and accepted assessment of experience, education and ethical standards.

This presentation gives an overview of the benefits of being licensed, the key players involved in becoming licensed and the path to licensure. The overview presents the background information to help put the other session presentations in perspective. More details about licensure are available in the March issue of *Chemical Engineering Progress (CEP).*

Value of being licensed

Being licensed shows other that you have a strong commitment to high technical and ethical standards. It can provide you with better career opportunities and provide a competitive edge when seeking employment. Many employers state that they would choose a licensed engineer over an unlicensed engineer, given both are otherwise equally qualified. Also, companies often offer incentives to encourage their employees to become licensed.

Some engineers think that licensure is important only in consulting or expert testimony. However, licensure is often required in firms doing design, construction and startup of processes. State laws also may include those teaching "advanced" engineering subjects be licensed.

Some states allow an "industrial exemption" for large companies who have the financial resources to cover liabilities arising from employee decisions. Most companies encourage licensure but many oppose eliminating the industrial exemption if they have operations in multiple states because it requires their engineers to be licensed in multiple states. There is a push to eliminate the industrial exemption. This important topic is discussed in another presentation.

No one can predict the exact future course of their career. However, being licensed may provide unexpected rewards and opportunities.

Key players in licensure

Becoming a P.E. involves education, experience and examination. Besides your coworkers, who can help guide your job experience, there are two important players in becoming licensed. The first is your local engineering board that establishes the standards for education and experience and ability to take the examinations. You will work with the board throughout the licensure process.

The second important player is the National Council of Examiners for Engineering and Surveying (NCEES) (www.ncees.org) . This is a nonprofit organization whose members are the various engineering licensing boards. The NCEES was created to provide some uniformity of standards of education, experience and examination. The membership establishes Model Law to offer uniformity. However, Model Law is a set of suggested best practices and no board follows Model Law exactly. All exams are administered and graded by the NCEES.

The licensure process

Obtaining a P.E. involves the following steps:

1. Having the appropriate formal education
2. Passing the Fundamentals of Engineering (FE) exam
3. Obtaining the appropriate experience (engineering internship)
4. Providing required documentation to board
5. Passing the Chemical Professional Engineering (PE) exam

While the general requirements are uniform across all jurisdictions, exact details vary. It is critical to work with your local board so that you understand and meet the exact requirements. A brief synopsis of each step is given below. More details are provided in the *CEP* article and in other presentations in this session.

1. Formal education: All jurisdictions require BS engineering degree, or equivalent, from an EAC/ABET-accredited four-year engineering program. Foreign degrees and other degrees are considered but must be approved by the board.
   Current Model Law states that as of 2020, engineers taking the PE exam must have a Masters or equivalent (MOE). This requirement is strongly opposed by AIChE and many other organizations. More details are presented in another presentation in this session and in the *CEP* article.

2. FE Exam: This exam tests your fundamental knowledge of engineering. It is easiest to pass when you are first eligible to take it (usually senior year or shortly after graduation). Another presentation and the *CEP* article provide more details of the exam and the upcoming changes in the exam.

3. Practical experience: Your work experience must be appropriate engineering practice. Typically four years are required although some boards permit fewer. In some jurisdictions advanced degrees can reduce the number of years required.

4. Documentation: The exact format of documentation and what is required varies between jurisdictions so it critical to work with your local board on this. However, all will require proof of education, passing the FE exam and work experience. The work experience will be supported by several references, who will also vouch for your ethical behavior.

5. PE Exam: This is the final step in becoming a licensed professional engineer. The exam is designed to be passed by a "minimally competent engineer with four years of practical experience". This exam tests the practical application of your engineering knowledge. Both a session presentation and the *CEP* article provide more details on the exam.

Summary

The AIChE recommends that its members become licensed to help ensure a competent and ethical chemical engineering workforce. While the process may seem daunting at first, working with your fellow employees and your local board will make the effort much easier. Each year about 400 chemical engineers become P.E's. They have made a significant investment in themselves and possibly in their future career.

(78b) Licensure--An Academic Perspective

S. Ranil Wickramasinghe, Department of Chemical Engineering, University of Arkansas, Fayetteville, AR

Most Chemical Engineering faculty members are not licensed. Why? Is it important for faculty to be licensed? Most undergraduates take the Fundamentals of Engineering exam but do not take the Professional Engineering exam. Why? Should faculty be licensed? Is licensure important for faculty who are conducting research? Should we encourage faculty to obtain their PE license? In this presentation these questions will be discussed.

TESTING FOR COMPETENCY - THE PROFESSIONAL ENGINEERING EXAMINATION

Anne Bertelsmann, PhD, PE, Galveston Bay Refinery, Marathon Petroleum Company LP, Texas City, TX
And
Denise Chastain-Knight, PE, CFSE, CH2M HILL, Augusta, GA

Introduction

The decision to pursue engineering licensure is a personal one, and the road includes many hurdles to overcome. Engineers take this road for a variety of reasons, with different motivations and experiences along the way, and finish the trek with the PE Exam. The Principals and Practice of Engineering Examination (PE Exam) is a nationwide exam used to ensure that licensed PE's are "minimally competent". The exam must include test content applicable to the competency being measured, effectively differentiate between competent and not competent candidates, and provide a consistent competency measure from administration to administration. This paper will define minimal competency, discuss the purpose of the Professional Engineering Exam, exam format using the Chemical Engineering Exam as the example, development methodology, item quality metrics and offer suggestions for test taking preparation. It will also describe the function of the NCEES Exam Development Committee and their role in developing a fair, effective exam, and explain why it takes weeks to get a result for a computer scored exam.

Path to Licensure

Whether an engineer undertakes the path to licensure early or late in their career, each must successfully negotiate the same hurdles. Each step in the process has a specific goal to either develop or to measure the engineer's capabilities toward professional competency. Figure 1 on the next page illustrates the path to licensure.

The quest to licensure begins with obtaining a Chemical Engineering degree. The next step is to pass an exam that measures competency on engineering basics. This first exam is the Fundamentals of Engineering Exam (FE) – also sometimes called the EIT (Engineer in Training) Exam. The FE exam is typically taken either during the senior year in college or shortly after graduation. The types of items resemble "textbook" questions and are designed to test understanding of the engineering fundamentals as taught in a typical engineering degree program. Currently, the FE examination consists of a general morning session (common to all disciplines) and a choice of seven discipline specific afternoon modules. Starting in the beginning of 2014, the FE examination will convert to a computer based test, which is discipline specific (i.e. no general module). For more information on the FE exam content and the conversion to computer based testing, refer to the NCEES website (http://ncees.org).

Figure 1: Path to PE Licensure

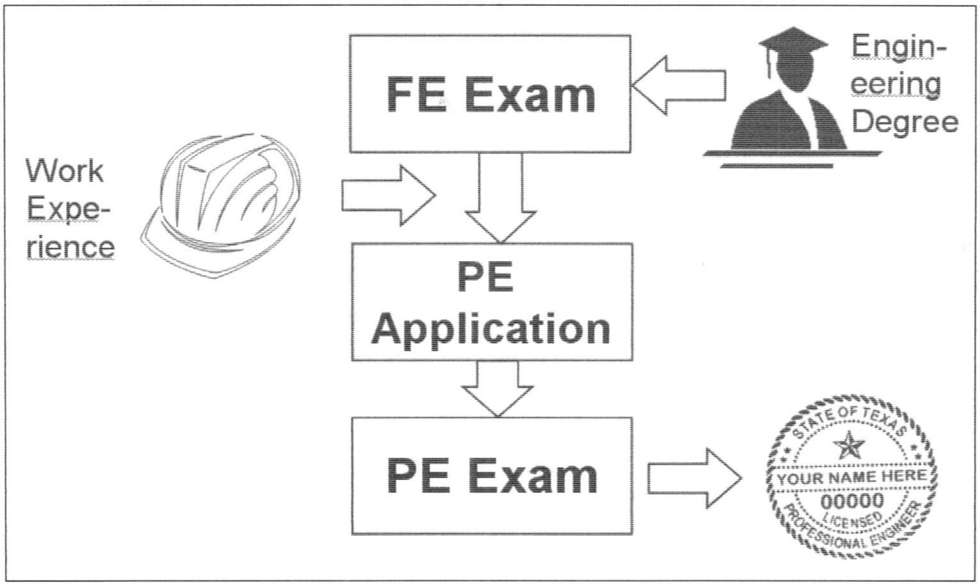

Following graduation, engineers enter the practice phase of their careers where fundamental knowledge is applied to real world application. In this phase of their careers, engineers use the theoretical principals learned from their education to solve problems that ultimately deliver physical products to market. Most states require a minimum of 4 years engineering related work experience before allowing candidates to sit for the PE exam. In order to become licensed, the candidate needs to submit an application to the state licensing board, describing education and experience, documenting passing of FE exam, and providing character and skills performance references. Format of the application and requirements for references, transcripts etc. vary from state to state. Once the application is accepted by the board, the candidate can register for the PE exam. After passing the PE examination, the engineer is licensed and can obtain the PE stamp. It is through this process that the state licensing boards ensure that only qualified candidates become licensed.

PE Exam Purpose

The PE exam is the final step in the road to licensure and its purpose is to ensure that only competent engineers become licensed. This helps to uphold the high standards of licensure and protects the health and welfare of the public. Since the exam is the same nationwide, it ensures consistency between states. Thus, one of the main requirements for a PE examination is that it can distinguish between competence and incompetence and that it can demonstrate this reliably and consistently. It is important to remember that "competent" should not be equated with "expert". The purpose of licensure is to identify individuals who have at least the minimum knowledge at which a person can practice professional engineering in such a manner that will safeguard life, health and property and promote the public welfare. That is why the term "minimally competent" is used.

A minimally competent engineer:

- should be able to perform identified tasks
- should meet minimum education and experience requirements
- is beginning to specialize in a sub-discipline
- may safely practice in another sub-discipline even though he/she does not practice in this area on a daily basis

The definition of "minimally competent" has implications on how the questions on the exam are written. The PE exam differs from the FE exam in that it addresses both the principles and the practice of engineering. So, while the FE questions are typically textbook related, questions on the PE exam are predominantly practice related. The questions build upon the experience that an engineer has gained during his or her first few years of working and need to recognize that chemical engineering experience can be gained in a variety of areas. The difficulty level of the questions is geared towards an engineer with about four years of experience.

PE Exam Content

The content of the examination is determined through a survey of practicing chemical engineers (PAKS study = Professional Activities, Knowledge and Skills), which is repeated about every 6-8 years to ensure that the examination stays current. The PAKS process includes conducting a technical content survey with input from practicing professionals. This survey is conducted over a period of weeks to ensure that the respondent pool size is adequate and is sufficiently diverse to be representative of the overall profession. Survey results are analyzed by a team of licensed Chemical Engineers, most with no prior exposure to PE exam development work, under the direction of a psychometrician who ensures that proper procedures are followed. The PAKS team will determine both the domain (subject area) content and item weight of the subdomains. PE exam content is established based on the PAKS study, and is published in the format of the PE exam specification available from the NCEES website (http://ncees.org/exams/pe-exam/) and is summarized in Table 1 on the next page.

The ChE exam specification was recently updated and became effective for the April 2011 exam administration. It includes six major domains and fifteen sub-domains that investigate theoretical and practical knowledge. Exam content is distributed within the major and subdomains according to the percentages shown in Table 1 to ensure a consistent content basis for every exam. The specification also includes guidelines for topical content within each sub-domain. For example the exam will include 4% of the items from sub-domain VI. C. Operation. Figure 2 includes example content from the 2011 exam specification.

Table 1: April 2011 ChE PE Exam Specification Summary

Knowledge Area	Approximate Percentage of Examination
I. Mass/Energy Balances and Thermodynamics	**23%**
A. Mass Balances	10%
B. Energy Balances and Thermodynamics	13%
II. Heat Transfer	**16%**
A. Mechanisms	9%
B. Applications	7%
III. Kinetics	**11%**
A. Reaction Parameters	6%
B. Reactors	5%
IV. Fluids	**16%**
A. Mechanical-Energy Balance	12%
B. Flow and Pressure Measurement Techniques	4%
V. Mass Transfer	**14%**
A. Phase Equilibria	5%
B. Continuous Vapor-Liquid Contactors	8%
C. Miscellaneous Mass Transfer Processes	1%
VI. Plant Design and Operation	**20%**
A. Economic Considerations	1%
B. Design	10%
C. Operation	4%
D. Safety, Health, and Environment	5%

Figure 2: Sample Exam Specification Sub-domain Content

C. Operation 4%

1. Process and equipment reliability (e.g., testing, preventive maintenance, startup/shutdown procedures, robustness)
2. Process improvement and troubleshooting (e.g., debottlenecking, experimental design and evaluation, optimization)

PE Exam Format:

The PE examination is given twice a year - once in April and once in October. It is an eight hour exam consisting of a four hour morning session and a four hour afternoon session with a lunch break in between. There are a total of 80 multiple choice questions (40 in the morning, 40 in the afternoon). Each question has four answer choices of which one and only one is the correct answer. Each question is stand-alone, i.e. it is independent of the other questions on the exam and can be solved with the information given in the problem, sometimes with the help of standard references. Some questions involve calculations while others are qualitative. All questions count equally and there is no penalty for guessing.

Each of the 80 multiple choice questions on the exam is of the same general format. In exam development, questions are referred to as "items" rather than questions because not every item has a question mark (like the one in the example below). Figure 3 illustrates the anatomy of an item, i.e. its general format and identifies the different parts of the item.

Figure 3: Anatomy of an Item

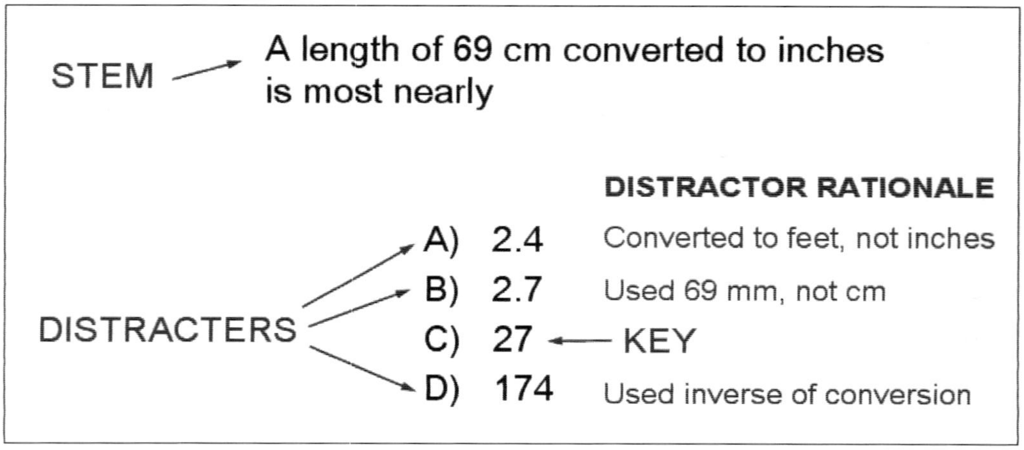

Each item consists of a stem and four answer choices. The stem (or problem statement) identifies what calculations are needed. It includes the information that is provided to solve the problem (sometimes even in table format) and states what is required for an answer. Beneath the stem are the four answer choices. Of these four answer choices, one (and only one) is the correct answer – the key. The other three are so-called distracters. They are incorrect answers – but they are not random numbers.

Each of the wrong choices represents a typical error that a less than minimally competent engineer (or someone not familiar with this specific topic) may make. Each distracter has to have a "rationale", i.e. an explanation which error was made. So, when a candidate is taking the exam and calculating a number that is exactly one of the given answer choices it does not necessarily mean that the problem was solved correctly. Maybe the candidate just made one of the "expected likely errors"...

The distracters can have a significant impact on the statistics of an item. If one of the distracters is obviously wrong (such as a negative absolute pressure), then even less than competent candidates will disregard it. That reduces the feasible choices, improves the "guessing" performance, and increases the chance that a not minimally competent candidate gets the question right. The Exam Development Committee strives to develop items that include the key, and three equally likely errors as distracters.

Exam Development

The PE Examination (as well as the FE, and the Surveying Exams) is a service that NCEES provides to the licensing boards. As such, NCEES provides the framework and the oversight for the exams, but the responsibility for the content rests with each exam committee. The committee writes the items, assembles the exams and reviews the statistics. NCEES has established rules and procedures that the exam committees follow to ensure consistency. Committees report back to NCEES with their exam quality measures. NCEES also provides the logistical support (psychometric consultant, publishing and printing, administrative support, maintaining records and the item bank, administering the exam, scoring, etc) and provides all security functions. For the larger exams, NCEES also pays and arranges for volunteer travel. For the smaller exams, this is the responsibility of the sponsoring society.

The Chemical Exam Committee is the group that actually writes and reviews the Chemical PE Exam. To be on the committee, one must be a licensed PE and express interest through the NCEES website or through a committee member. The committee has about 50 active members and meets twice a year in Clemson, South Carolina at the NCEES headquarters. Not everyone comes to every meeting, of course, so typically 30 – 35 committee members present at any given meeting. During the meeting items are written and reviewed by committee members who work with one another to critique and improve the items. Task teams also work to assemble, review and pretest exams.

Before an exam is administered, each item will have no less than six documented formal reviews and likely have been assessed by more than 20 people. Committee members are asked to complete detail oriented and often challenging engineering work and can earn PDH (professional development hours) for this activity. Developing the PE exam helps committee members to keep their engineering skills current and to give back to the profession. Without the work of this committee, the PE exam and with it the licensing wouldn't happen.

Preparing an exam that effectively and consistently measures minimal competency is a challenging task that succeeds or fails based on the ability of the committee to predict item performance. Each committee relies on a structured process and employees a variety of quality tools and techniques to improve the opportunity for success. Figure 4 is a workflow of the exam development activities. Items are written by a PE and submitted for review. Many items are prepared by the committee members at each meeting, while others are submitted by PEs who do not attend the meetings. The individuals who write and review items are referred to as Subject Matter Experts (SME). The author and two reviewers will independently work the item and develop the rationale for each distracter. Through this process the item is often revised and edited to improve its presentation and quality. This step is considered one of the most critical activities of the committee because items are the building blocks of the exam and each must meet high quality standards. Each committee member is responsible for item quality and is granted authority to modify draft items as needed to meet quality objectives, or reject items that fail criteria and cannot be edited to usable items. Finally the item is given a final check by the committee chair (or authorized designee) to verify that documentation is complete, both reviewers agree to the item content, and all quality checks have been completed, and approves the item for acceptance into the item bank. The item bank is a repository for items that are ready to be placed on exams.

Figure 4: Exam Development Process

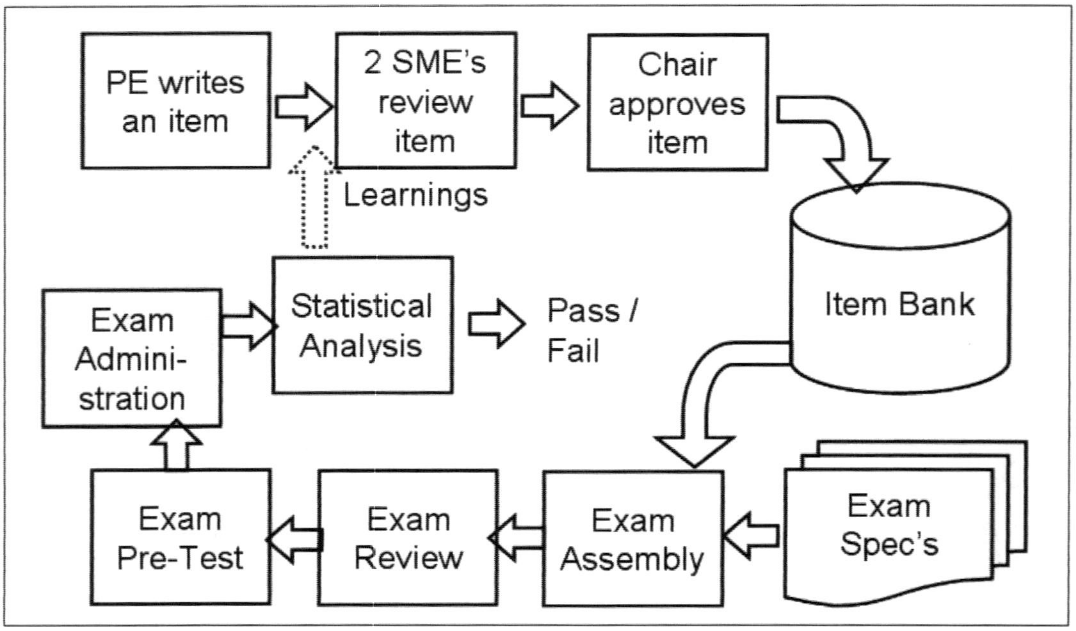

Exams are built, reviewed, revised, reviewed, quality tested and reviewed over a series of meetings. Each exam will have three independent team reviews, a pretest and a quality assessment. There are specific checks required at each workflow gate with task teams assigned to maximize the number of eyes that evaluate any given exam. An exam is compiled using items from the bank following the content distribution of the specification and sequenced in the specification order. The build team will then conduct a complete review of the exam, looking at each exam item for individual quality and checking for overall content flow and replacing items if necessary. The exam is then reviewed in two more meetings by independent teams who provide a "fresh set of eyes" and continue to improve the content to ensure that each item on the exam and the overall exam meet quality criteria. They also verify that the documentation is complete in accordance with the NCEES exam development requirements. Each exam must also have a pre-test where a group of committee members take the exam in as if they were actual test takers. After working all the items on the exam, they answer a series of questions about the items designed to predict item performance, then perform a distracter evaluation. Comments from the pre-testers are incorporated into the exam. In the final meeting before administration, a quality team conducts one last review of the exam verifying that all preceding checks (and associated documentation) are complete, all comments have been addressed, specification distribution is correct, and that author diversity limits and other NCEES criteria are met.

When the exam is administered and scored, data is compiled that can be used to evaluate the quality of the exam as a whole, and individual items. The data is analyzed and used to help the committee to learn what makes items perform effectively, and improve our performance prediction skills. Authors are provided a 'report card' so they know how items they wrote performed and have the opportunity to see how those items evolved in the review process. At each meeting, the committee will review items that had great statistics and try to find out what made them perform well. Similarly, we will review items that didn't perform well and identify traits that are undesirable in future items.

Item Statistics

Item quality is ultimately measured by the statistics generated from the candidate responses after an exam administration. Before discussing the statistical analysis of the examinee responses, one must first consider what makes an item a "good" item. The definition of a good item is one that is capable of distinguishing between "minimally competent" and not "minimally competent" candidates, does not discriminate against any particular group of test takers, and performs consistently in multiple exam administrations. To achieve those goals, it must be relevant, not confusing, and measure competency.

Most importantly, the content of the item must be relevant. Relevance is defined as the appropriateness of the topic and the difficulty of the item. The topic of the item must test knowledge or skills that are directly related to the exam specifications. For example: a question on high voltage power distribution may be relevant for an electrical engineer, but will not provide information on engineering competency of a *chemical* engineering candidate – no matter how well written it is. If the item does not relate to the exam specifications, it does not belong on the exam. Second, the difficulty level and complexity of the item must be such that a "minimally competent" engineer with 4 years experience can solve it in about 6 minutes. When writing items, authors often think about the interesting and challenging problems that are encountered in practice; however, if an engineer with 20 years experience requires several hours or days to solve the problem, what are the chances that a junior engineer will get the answer in 6 minutes? These type of problems, as interesting as they may be, do not make good exam items.

Another important principle to consider is that items should not be confusing to people who know the answer. When an item is confusing, a competent engineer who understands the concept being tested may not select the key, thus the question does not help to distinguish between competent and not competent. The so called "trick" question does not evaluate a candidate's competence and is considered a "bad" item to be weeded out in the exam development process.

It is equally undesirable to use items that give the answer away to people who don't know the answer. If the item includes clues to the key, then an incompetent engineer can choose the right answer without understanding the subject matter; thus the item does not measure competency. Clues can take many different forms:

- If some of the answer choices can be ruled out as impossible (like negative absolute pressures), the chance of guessing the right answer goes up.
- Grammar can give clues when one of the answer choices "stands out", i.e. is more detailed or does / does not match the grammatical structure of the item statement

For example, an item that asks about the 19th president of the United States, where three of the four answer choices are well known pop stars, does not require the candidate to know anything about US history to pick the correct answer.

Following an exam administration, a psychometric analysis is performed (psycho – metric = measuring the mind) on examinee responses. The analysis is solely based on information provided on the marked answer sheets without regard to the actual content of each item, but quite a bit of information about an item can be observed just from examinee responses. The psychometric analysis will evaluate a number of things. Of course, the simplest analysis is to determine how many candidates selected the key. Statistically, there is a 25% chance of guessing the correct answer because there are four choices. An item where less than 25% of the candidates selected the key may be flawed because it may have directed candidates towards preferentially selecting a distracter over the key. The next question is to ask *who* got the question right. Ideally, the candidates who got the question right were the same people who performed well on the exam overall. The analysis looks at the distribution of answers between the distracters and who selected them. This analysis also considers if all distracters were selected equally or if there was one distracter that was particularly attractive; maybe it even got selected more frequently than the key.

The objective of this analysis is to confirm the validity of each item as a tool capable of measuring engineering competence. An item that follows the basic three criteria (relevance, no confusion and no clues) can be expected to distinguish between competence and incompetence; however, sometimes clues, confusion or complexity can creep into an item unintentionally and be missed during the review before the exam is given. Statistics are then used to identify potential flaws. One such flaw is a mis-key. Mis-key is when the committee identified for example option A as the correct right answer, but option B is actually correct. Another flaw is an ambiguous item where more than one of the options could be correct. Despite the committee's efforts, such flawed items occasionally make it through the review process and are placed on an exam. The rigor of review process prior to exam administration minimizes the likelihood of a flawed item; however, it cannot totally eliminate them. The statistics after the exam will reflect item imperfections and provide an opportunity to compensate for item flaws in exam scoring so that candidates are not negatively impacted by flawed items.

Figure 5 is an example of the statistical analysis report for an item prepared by the psychometric consultant. One such sheet is prepared for each of the 80 items on an exam. The top left corner shows the "number of cases", i.e. the number of examinees who took the exam in this administration. In this example, there were 2910 candidates. So, this example is not from any Chemical PE exam where typically about 300 candidates sit per exam. The number below the case number (highlighted in yellow) is the Mean (or the P+ value). This number shows what percentage of the candidates selected the key; in this case, 65%. P+ alone indicates that the question is of appropriate difficulty for the average test taker; not too easy and not too hard. Statistically, the higher the P+, the easier the item.

Figure 5: Item Statistics

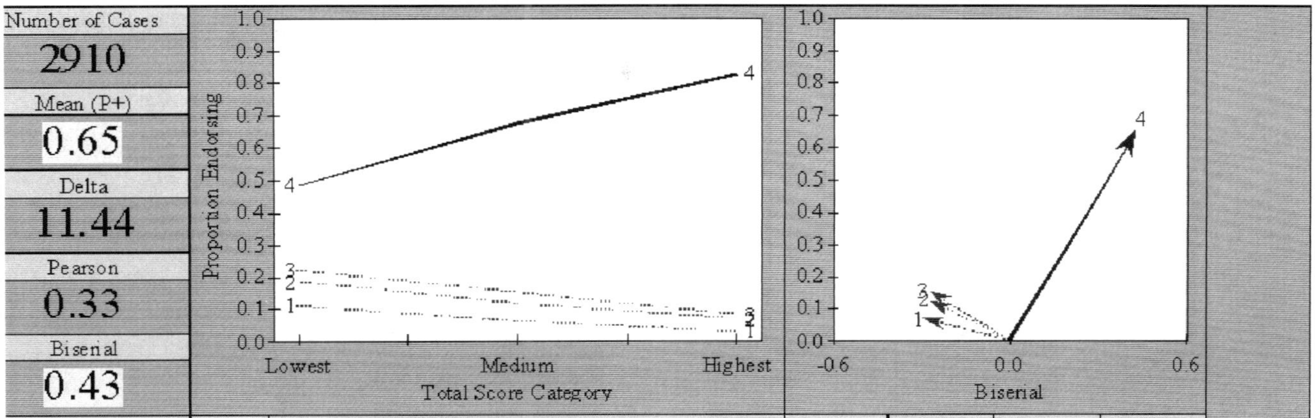

Alt.	Lowest	Low	Medium	High	Highest	%	N	Mean Scr	Biserial
1	0.11		0.06		0.03	6.7%	195	43.0	-0.29
2	0.19		0.12		0.07	12.5%	365	44.1	-0.27
3	0.22		0.15		0.08	15.4%	449	44.3	-0.27
4*	0.48		0.67		0.83	65.2%	1896	50.1	0.43
OM	0.00		0.00		0.00	0.2%	5	42.4	-0.20
Mn Scr	38.68		48.50		58.27				
Grp N	1025		1022		863				

The statistics not only evaluate how many candidates selected the key, but how selecting the key on this item correlates to overall exam scores. Test takers are divided into three groups of roughly equal size based on overall score on the exam; the low scoring third, the medium scoring third and the high scoring third. For each of these groups, the report in the table under the left graph shows how many selected which answer choice. For example on this question, "4" (or "D") was the key as indicated by the asterisk. Of the low scoring third, only 48% got the right answer, while 83% of the high performers selected the key. Thus, someone who performs well on the exam overall (which is an indication of competence) is significantly more likely to select the key than someone who scored low (may indicate lack of minimal competence). This spread shows that the question is capable of testing competency.

Figure 5 also illustrates that the incorrect answers were relatively evenly distributed between the distracters, indicating that none of the distracters appears to be overly attractive or easily ruled out. Another way of looking at the same data is presented on the right of the page. The table shows average overall scores for the people who selected each of the answer choices. In this case, the people who selected the key had an average overall score of 50, while the people who selected the distracters had overall scores in the range of 43 to 44. Five candidates with an average score of 42 did not answer the item. This clearly demonstrates that the people who selected the key were more likely to do well on the exam overall, and indicates that the item can effectively measure competency.

The data is evaluated in yet another way by the cumulative statistics in the lower left: Mean of Passers indicates that of those candidates who passed the exam, 78% selected the key. Mean of Failers shows that only 53% of the people who failed the exam selected the key. The larger the spread between these two numbers, the better the item is at measuring competency.

A handy way of summarizing the "who got the question right" analysis is the R-Biserial number. Basically, this number is related to the slope of the lines in the graph on the left. For a good item, this slope for the key should be positive and the larger the better because it expresses the difference in performance of the low and the high scorers. Theoretical range for the R-Biserial number is from -1 to +1, positive values indicate that the candidates who selected the key also scored higher on the overall exam. A negative R-Biserial number indicates that higher scoring candidates selected one of the distracters, which could be an indication of a mis-key or ambiguous item. The R-Biserial number is highlighted in yellow on the left and shown in the picture on the right. The R-Biserial number is important because it evaluates how well an item meets the three basic criteria. We previously observed that the sample item is of appropriate difficulty based on the percentage of correct answers (P+ value). The R-Biserial number shows that we did not confuse the competent candidates (a high percentage of high performers got it right) and apparently it did not give any clues to the less competent candidates (. It also appears to be relevant with regard to the topic since the high performers (those with better understanding of chemical engineering) did better on this than the low performers.

To further illustrate the concept of the R-Biserial number, consider an example of an item that might produce undesirable statistics. The item is asked on the final exam for a graduating class of chemical engineers: "Which bar in town has the most beers on tap?" One would expect a certain percentage of students to select the correct answer, but are those candidates the most knowledgeable students who spend a lot of time in the lab, or the "party crowd"? This question would likely have a negative R-Biserial number, indicating the overall least knowledgeable students are more likely to select the key, which means this item is not an effective measure of a graduating chemical engineer competency, nor is it relevant to chemical engineering knowledge and skills.

Determining Competency

The individual item statistics are the building blocks for determining the overall exam performance. To make the move from the individual item statistics to determining competency of the candidate, a passing score, also called cut score, needs to be established. Figure 6 on the next page illustrates the philosophy behind relating the score distribution to the cut point, or passing score. Total score is simply the number of items answered correctly. For the Chemical Engineering PE Exam, the total score has a theoretical range of 0 to 80. Most candidates have at least some knowledge and get lucky guessing on some questions, so scores on the very low end are uncommon.

If it were known upfront which candidates are competent and which ones are not, the scores of the two groups could be plotted as shown in Figure 6. The at least minimally competent engineers have higher scores, the not minimally competent engineers have lower scores. However, there will be some overlap between the two groups. The better individual items differentiate between competency and incompetency, the smaller the overlap. However, because there is a range of competency and because a person's performance can vary, there will always be some overlap and thus the cut point decision is not clear cut.

Figure 6: Score Distribution

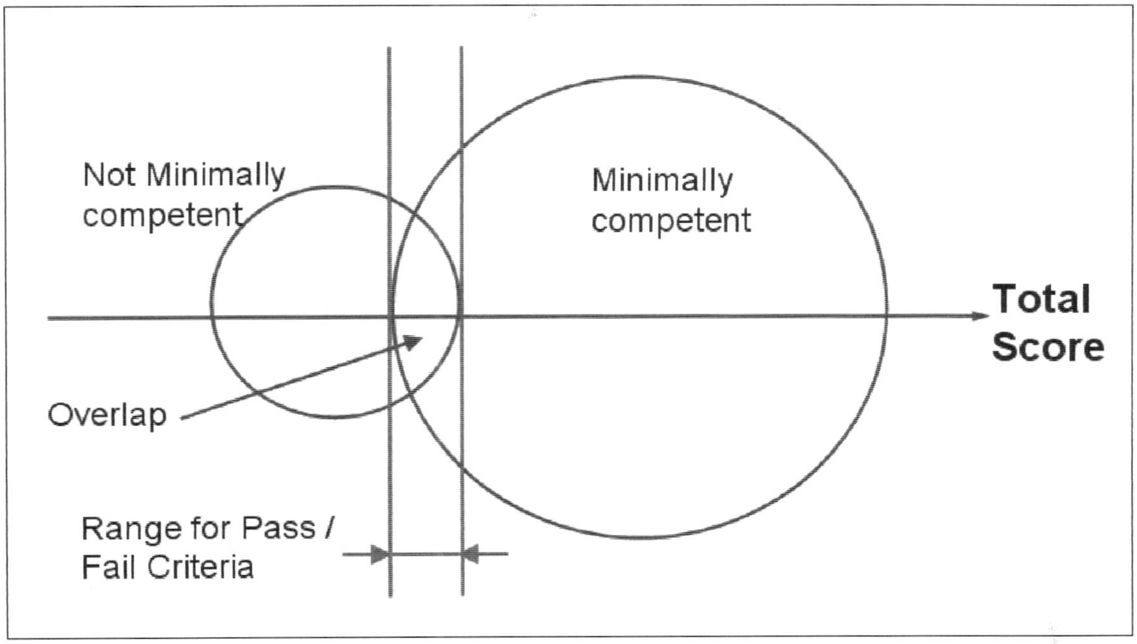

The overlap represents the range where the pass/fail decision should be made. All the at least minimally competent candidates on the right side should pass, all the not minimally competent candidates on the left should fail. With any decision point within that overlap range, there is a potential for competent engineers to fail and/or for less than minimally competent engineers to pass. Failing a competent engineer means that they will need to re-take the exam and includes the risk that they may give up on licensure. Passing a less than minimally competent engineer may allow an under-qualified person into public practice. Since the second type of error is more consequential, there is a tendency towards raising the bar when determining the pass/fail criteria.

Without knowing upfront which candidates are "minimally competent' and which ones are not (after all, it is the purpose of this exam to establish this), statistical measure are used to determine the pass/fail criteria, i.e. establish the "cut score". The cut score is the number of correct answers that a candidate needs to have in order to be considered "minimally competent". One way of determining the cut score is through panel of licensed PE's. This process is fairly involved, since it requires about 15-20 PE's who can volunteer 2-3 days of their time and travel to attend to cut score session. The panel must have good diversity (age, gender, area of practice, geographic location, demographics, etc…) and the cut score study must be completed before the results of the exam can be reported back to the state boards and ultimately to the candidates who are eagerly waiting for the scores. For these reasons, a cut score panel is usually only done when there was a change in exam specifications where the exam results cannot be directly compared to recent prior exams. The first exam where the format or specification is changed is called the "anchor exam" and cut point is established by a formal study.

For a cut score study the diverse panel of volunteers (all licensed PE's) meet and spend about a day to work the exam in a timed environment just like the actual administration. They can use any references, cannot consult with one another and they don't have access to the exam solution. The next day, they receive their score, and review and discuss each item as a group. They look at the statistics similar to those described above, but lacking the pass/fail metrics, review the solution, the distracters and discuss the topic and difficulty of each item. On the basis of that discussion, the panel members make a judgment call on how many "minimally competent" engineers out of 100 would answer this question correctly. A psychometrician guides the session and collects these estimates from each panel member. This data is then used in statistical analysis, generating a proposed passing score. The proposed passing score can then be adjusted (for example, to err on the "safe side"). This then becomes the basis for the pass/fail decision for the anchor exam.

Such a cut score study is not practical for every exam at every administration. So as long as there is no change in the exam specification, it is possible to "equate" the current exam to a previous exam. The equating process, heavily supported by a psychometrician, compares examinee performance on the current exam with examinee performance on previous exams. From this comparison it can be determined if the current exam is harder or easier than the previous exams, and cut point adjusted accordingly. If the data indicates that the exam is easier, the cut score is adjusted upward and a candidate will need to answer more questions correctly in order to pass. Conversely, candidates are not penalized if they take a harder exam because the cut point is revised downward and fewer questions must be answered correctly to pass. Equating is NOT the same as grading on a curve. When grading on a curve, the passing percentage would be fixed and the cut score adjusted to achieve that percentage. With equating, the competency of a candidate is measured on personal performance, and not dependent on the individual candidate's position within the test taking population, nor on the difficulty of the exam.

The psychometricians calculate other measures from the data. Coefficient Alpha measures the internal consistency or statistical homogeneity of the data, thus basically expresses the reliability of the exam. A measure called "Subkoviak" provides an estimate of the consistency of the pass/fail decision or an estimate of the probability that an individual would receive the same pass/fail decision on multiple administrations of the exam. All these measures are regularly reviewed to ensure that the exams can consistently measure competency.

After an exam is administered and all the answer sheets are scored, the statistical item analysis is performed. Exam Committees need to review potentially flawed items in case the key needs to be adjusted and the exam rescored. Then the cut point is determined, either through a cut score study with a panel or through equating. Only then can the pass / fail decision be made. State boards are notified with the results, perform state specific pass/fail assessments and then notify candidates of the results. NCEES recommends that only pass/fail should be reported to the candidates, though some states provide the actual score (or a "scaled percentage" based on a fixed passing percentage) to candidates. Candidates who fail get a diagnostics report so that they know how they performed in the various subject areas. NCEES recommends to only report the pass/fail decision for two reasons: The cut score can vary from exam to exam, so a candidate may pass one year and another candidate taking the exam another year may fail with the same score. The other reason is that the exam is designed to make a pass/fail decision on minimum competency. In that regard, the PE exam is different from an exam like the SAT, which is designed to measure "aptitude" over a wide range of abilities. The PE exam is not designed to measure ability accurately at the far ends of the scale. The PE exam does not distinguish between the "outstandingly competent" and the merely "extraordinarily competent engineer". Similarly, it cannot distinguish between the "totally incompetent" and the "absolutely clueless". Thus, a candidate who received a score of 75 out of 80 is not necessarily a better engineer than a candidate with a score of 72 out of 80 – both are "minimally competent" and pass.

The skeptical reader asks; "Does this really work?" Can an exam such as this really measure competency as the statistical "number magic" appears to indicate? The answer is found in the historical exam pass rates. NCEES does not report an overall pass rate for the exam, but breaks it up into a pass rate for first time takers and for repeat takers. Table 2 includes pass rates for exams given under the current Chemical PE Specification. The most recent pass rates for other exams can be found on the NCEES website.

One specific item to note about the data in Table 2 is that there is a gap between the pass rate for first time takers and for repeat takers. The pass rate for first time takers is typically rather high; in the 70 – 80% range for the Chemical PE exam. This high pass rate is because the examinee pool is already a pre-selected group. In order to be admitted to the exam, a candidate had to pass the FE exam, had to prepare an application with documentation of engineering experience, and had to be approved by the state board, a process that weeds out some of the less qualified potential candidates. Also, especially in chemical engineering where there are many jobs that come under the industrial exemption and do not require licensure, the people who decide to pursue licensure are already a non-random subset of all engineering graduates. This contributes to the relatively high pass rates for first time takers. The portion of the candidate population that passes on a repeat attempt includes in candidates that are minimally competent but may have been inadequately prepare in their first exam sitting.

Table 2: Chemical PE Exam Pass Rates

Chemical PE Exam Pass Rates	**April 2011**	**October 2011**	**April 2012**	**October 2012**
1st Time Pass Rate	78%	75%	74%	71%
Repeat Pass Rate	31%	29%	31%	20%

How to Prepare for the Exam

All the fascinating information presented in this paper aside, chances are the reader really wants to know how to prepare for the exam. Preparation philosophy will differ for a young candidate near the 4 year experience mark from that for a veteran engineer. One theory is that any minimally competent engineer can pass the examination without any preparation if he or she had unlimited time - time to look up references, read up on concepts and topics outside one's familiar area of practice, find the right equations and conversion factors, etc. But there is a time limit, and 6 minutes per question is not a lot of time.

Exam preparation should focus on building speed and efficiency for solving items. Through study and preparation, a candidate should become familiar with reference content, placing tabs to help find frequently used tables or compile a personal reference of frequently used information. Common equations and frequently used conversion factors should be memorized. Candidates should reactivate the habit of using math: solving equations, working with variables and using a scientific calculator instead of a computer. Speaking of calculators, it is strongly recommended to prepare for the exam using one of the approved models. Precious time can be wasted searching for the logarithm button on an unfamiliar calculator. Approved calculator models can be found on the NCEES website.

Exam readiness is achieved through lots and lots of practice. Preparation classes are useful for refreshing material basics, but are no substitute for rehearsal. A candidate should prepare by solving as many sample problems as one can find. By solving many, many problems, one will automatically remember the frequently used equations and conversion factors, enhance math skills and remember where to find information. The process also develops test day confidence because the items encountered on the exam have been encountered in similar form during exam preparation. Candidates should conduct one or more mock exam day(s) to practice pacing and practice maintaining focus for 8 hours. NCEES publishes a sample exam for each specification that has been developed by the exam committees in the same manner as a "real" exam. One author of this paper frequently recommends to potential exam candidates that they practice with readily available materials, complete a timed trial with the NCEES exam, then continue practicing with more materials.

Finally each candidate should prepare for the test environment and stage to be at peak performance on test day. Exams are administered in finite locations within a state and often require candidates to travel. Plan a travel itinerary that will allow arrival at the test site early, rested and unstressed. Have a satisfying but not heavy breakfast and make a plan for a relaxing lunch. Consider the test room conditions, dress comfortably and in layers. Be aware of the materials that are allowed in the test hall, and prepare accordingly. Candidates will be turned away without a valid government photo ID. Cell phones and other forbidden materials may be confiscated or admission refused. The exam is worked in a printed booklet, which contains sufficient space for working the questions and pencils are provided. Candidates need to bring an approved calculator, unapproved calculators will be rejected. The exam is open book, so candidates also need to bring any references that they want to use during the exam. References must be bound (no loose paper), but can be marked with flags. Review the test day policies on the NCEES website and plan accordingly (http://ncees.org/exams/exam-day-policies/).

Closing Comments

Professional Engineer Licensure is a personal decision and requires one to undertake a dedicated and challenging journey. Candidate should embark on this journey knowing that the exam development process is designed to ensure each exam measures for minimal competency in a consistent manner, and items are statistically validated. Finally, the factors influencing exam score processing timelines are understood. The authors of this paper encourage engineers to pursue licensure. We speak for ourselves and our fellow PEs when we say that nobody ever regrets the resources and effort spent obtaining a PE, and we appreciate the industry approval and instant expert acceptance granted by those two little letters following our names. P.E. Professional Engineer.

(78d) The Current Hot Button Issues in Domestic Professional Licensing

Amos Holt, Southwest Research Institute, San Antonio, TX

Amos Holt, a past president of ASME and current VP for Environmental, Safety and Quality Systems will review the issues currently receiving the greatest attention by the PE community which includes all of the various state and territorial licensing boards, the National Council for Examiners of Engineers and Surveyors as well as the numerous professional society stakeholders and educational organizations and societies. The discussuion will touch on the Model Law and its currect sections on the Masters or Equivalent (MOE) requirements as well as Industrial Exemptions. The various positions of key players covered including the position of a consortium of professional societies called "Licensing that Works" or LTW will be summarized. This group includes both ASME and AIChE and represents about 240,000 engineers thtroughout the world. Amos will tell you what you will likely be hearing more about in the next few years and how it might affect both current PE licensees and prospective PE licensees.

The California Divide – And Why You Need To Engage Your Legislators

Diane Spencer, PE, Lawrence Livermore National Laboratory, CA
and Emmett Miller, PE, consultant and AIChE Fellow, CA

Abstract

California is crippled by a great divide that has nothing to do with tectonic plates. California's professional engineering statute gives civil engineers free reign, allows electrical and mechanical engineers to work in their disciplines so long as that does not conflict with civil engineering, and acknowledges by title only that other engineering disciplines exist.

This division presents problems for engineers whether they have a professional engineering license or not. For example, Spill Prevention Control and Countermeasure (SPCC) plans are mandated by federal regulations. Such plans are prepared by chemical engineers all over the country, but NOT in California. SPCC plans are part of the exclusive domain of civil engineers.

A fire protection engineer was once told he could not design a fire suppression system because that was the exclusive domain of mechanical engineers. A chemical engineer was once told he could not submit a Risk Management Plan (RMP) (required by Cal EPA) without a civil engineer stamp because even the hazard analysis on a facility that uses chemicals is the civil engineer's domain. These are just a few examples where engineers are not allowed to work on projects that similarly trained engineers are expected to do in the rest of the country and the rest of the world.

So how do engineers in California navigate this regulatory nightmare? And how do we influence the legislators, engineering board, and others to nudge the statute in the right direction?

California's answer included forming a coalition of professional engineering societies that work together to present a united voice. The California Legislative Council of Professional Engineers (CLCPE) not only represents the interests of engineering professions to the legislature, but serves as a pro bono resource for the legislature on technical issues as well.

This paper will discuss: (1) our lessons learned from working with legislators; (2) California's professional engineering statute and why it is written as it is; and (3) why one needs to be engaged in the legislative process. Professional engineers are entrusted with protecting the health, safety, and welfare of the public. Getting engaged has the benefit of letting legislators understand the impact of their decisions and allows engineers to serve the communities where they live, and provide a safer, better community for everyone.

California's Problem

While most other states utilize some form of a generic registration model for engineers, California's model is discipline-specific, with civil engineering held on a pedestal. The definition of "civil engineering" in the Professional Engineers Act (California Business and Professions Code Sections 6700-6799.) is so broad that it encompasses nearly the entire spectrum of engineering. This overly broad definition, coupled with a mandate that only civil engineers can do civil engineering, means there is little room for those in other engineering disciplines to legally work in their own professions. The law allows overlap from civil engineering into all other disciplines, but other disciplines cannot overlap into civil engineering. So, not only are chemical engineers impacted, the other engineering disciplines are impacted as well.

Section 6731 states the following, with parts underlined for emphasis:

> "<u>Civil engineering embraces the following studies or activities in connection with fixed works</u> for irrigation, drainage, waterpower, water supply, flood control, inland waterways, harbors, municipal improvements, railroads, highways, tunnels, airports and airways, purification of water, sewerage, refuse disposal, foundations, grading, <u>framed and homogeneous structures</u>, buildings, or bridges:
>
> (a) The economics of, the use and design of, materials of construction and the determination of their physical qualities.
>
> (b) The supervision of the construction of engineering structures.
>
> (c) <u>The investigation of the laws, phenomena and forces of nature</u>.
>
> (d) Appraisals or valuations.
>
> (e) <u>The preparation or submission of designs, plans and specifications and engineering reports.</u>
>
> (f) <u>Coordination of the work of professional, technical, or special consultants</u>.
>
> (g) Creation, preparation, or modification of electronic or computerized data in the performance of the activities described in subdivisions (a) through (f)."

For many years the conventional wisdom has been that the term "fixed works," as used in the statute, (see Section 6731 above) somehow limits the scope of civil engineering, leaving room for other engineers to practice their craft. In order to clarify where that boundary lies, Senator Mimi Walters asked the California Legislative Counsel Bureau for a legal opinion (2011). The findings were shocking. <u>"Fixed works" includes not only the structures, but also all the attachments to the structure.</u> So, if a mechanical engineer designs a pumping system or a chemical engineer designs a fractionating column, and that hardware and the hardware is secured to a foundation, those engineers have crossed into the exclusive domain of the civil engineer. Other engineers are similarly impacted.

When a reasonable person reviews this list of studies or activities, it becomes apparent that the list precludes even scientists from legally studying the phenomena and forces of nature. And in most jobs, engineers are expected to produce engineering reports or to coordinate the work of other professionals.

How Did California Get Into This PE Mess?

The perverse nature of the current Professional Engineers Act in California is a result of good intentions gone awry.

In March 1928, a dam northeast of Castaic, CA, broke and flooded the valley below, wiping out the town of Santa Paula and flooding other towns downstream. Hundreds of people died and millions of dollars' worth of property was damaged.

The Professional Engineers Act was then created to ensure that only qualified people (civil engineer PEs) could work on dams and other fixed works. The legislators made the law as inclusive as possible to make sure there were no loopholes. Later, the Legislature expanded the Act to include Chemical, Electrical, Mechanical, Petroleum, Industrial and Metallurgical Engineering as Title Act Disciplines. With the passage of time, the Board for Professional Engineers, Land Surveyors, and Geologists (BPELSG, aka "Engineering Board") was empowered by the Legislature to create Title Disciplines on its own authority, which it did, eventually increasing the count to 15 title protected disciplines (Title Act disciplines), These are defined by Board Rules (16CCR 400-476) and have minimal stature in law. Along the line, Mechanical and Electrical Engineering were converted by statute to practice protection (Practice Act discipline)[1].

This patch job to add mechanical engineering and electrical engineering to the Professional Engineers Act has been ineffective (per Legislative Counsel's review). Inclusion of Title Act disciplines constitutes an even lower tier where those engineers are not able to take responsible charge of work within their field of expertise. In summary, the field is far from level.

The problem with the Professional Engineers Act has been known for many years. The difference now is the Legislative Counsel's legal opinion that was issued in July 2011 (www.CLCPE.org). According to that opinion, companies that rely on engineering services are on notice that the only engineering discipline allowed to conduct engineering work (not just supervise) in California is civil engineering. If a project requires a chemical engineer, a dual-licensed engineer will be needed to meet the competency requirements of both a chemical engineer and the legal requirements of a civil engineer. A dual license is needed for the other engineering disciplines as well.

[1] Four of the title disciplines (Corrosion, Manufacturing, Quality and Safety) were dropped ~1999 for lack of an NCEES exam.

The Industrial Exemption

Common to most states, California has an Industrial Exemption that allows companies engaged in manufacturing to use unlicensed engineers in their businesses— except for civil engineering work. The justification for the Industrial Exemption has been that such employers are capable of and competent in evaluating their own engineering talent, that they do not need interference from the state in doing so, and that failure that will impact the public is subject to strict liability.

There are exemptions for people who perform some specific and narrow civil engineering services. Architects performing architecture and building contractors following building codes are examples. But no general exemption exists for engineers who are not civil engineers if they perform "civil engineering." The Industrial Exemption even states that it does not apply to civil engineering.

With the recent legal opinion, however, all engineering practice in connection with facilities that are geographically fixed is now regulated as civil engineering, and the Industrial Exemption is therefore meaningless.

Engaging the Legislature

In the late 1940's, engineering societies in California organized the California Legislative Council for Professional Engineers (CLCPE) in order to work together as a resource to assist the California legislature, support laws that benefit the public and profession, and help keep society members informed about legislation that might affect them. Application of the Professional Engineers Act was not a problem to engineers in the state so long as the Engineering Board and regulators maintained a liberal interpretation. Likewise, CLCPE did not intensify its effort to change the law until the literal interpretation ultimately reinforced by the Legislative Counsel ruling came into focus.

CLCPE's efforts to change the law began in the 1980s. Our members helped draft a bill and found a senator to carry it. In California, the legislator who carries the bill is the "author" and organizations that support the bill, such as CLCPE, are "sponsors" regardless of who actually drafts the bill. Drafting a bill and finding "authors" is a process CLCPE repeated about a dozen times.

Let's reframe the process we used in California in terms that may help others take advantage of our lessons learned.

1) Coalition. We had a consortium of engineering societies working together as a coalition. This provided more credibility in that we were able to speak with one voice instead of individual organizations promoting separate proposals. A coalition also allows a sponsoring organization such as CLCPE to represent a larger portion of a legislator's constituency. AIChE local section leaders met with a senator to promote the one bill that chemical engineers sponsored without going through CLCPE. The senator listened intently about the impact of current law, asked a few clarifying questions, and then asked about how many chemical engineers were in the state. She was not satisfied with our answer, stood up and thanked us for our time.

Creating a coalition of several engineering societies enabled us to afford a lobbyist. AIChE and local sections are a 501(c)(3) nonprofit organization, which restricts how much money can be spent on lobbying. CLCPE however, was organized as a 501(c)(6) nonprofit, which allows for lobbying, but since the societies and CLCPE are all volunteer organizations, we do not have an intense political presence, and that restricts campaigning for or donating to candidates or political parties. Unfortunately, the opposition we face in California has major campaign funding and they openly support candidates that vote as the opposition wants them to vote.

Another advantage of working within a coalition of engineering societies goes beyond promoting legislation. The conversations we had when we met allowed us to gain insight into the problems each of the different disciplines were facing and we heard personal testimony about those various issues. This perspective has been invaluable.

2) Hiring a lobbyist. Over its lifetime, CLCPE has successively employed three lobbyists who helped with efforts to change the Professional Engineers Act, in addition to supporting the overall mission. A good lobbyist helps focus the discussion and facilitate consensus on language for a given bill. They also know how to navigate through the legislature, explain the bill or other proposals to legislators, legislative staff, and to others, and to find legislators who will author a bill. Typically engineers have full-time jobs, and working a bill is very time-consuming. The first of CLCPE's lobbyists was an engineer who later became Executive Officer of the Engineering Board. The other two were professional lobbyists with legal backgrounds.

3) Language for the bill. Earlier this paper discussed drafting language for a bill and then entering that bill into the legislative process (i.e., finding an author). "Entering" is the key word here. In the late 1990s, our bill attracted attention from the Professional and Technical Consultants Association (PATCA). In our quest to expand the law to allow engineers to practice their own profession, we needed to be careful not to adversely impact scientists (e.g., chemists) and other technical professionals. After several meetings with PATCA, we were able to work out compatible language. Going forward on subsequent bills, we deliberately engaged with the chemists (PATCA) early in the process.

4) Bill's "author". Over the decades we have been pursuing change, our bills have been authored by different senators and assemblymen. In several cases, the legislator had been an engineer in his/her former career. One such senator, who had been an aeronautical engineer, had commented that it took multiple disciplines to design an aircraft. A legislator with a technical or science background readily grasped the issue we were dealing with. One of the first committees to hear a bill dealing with engineering (i.e., the bills we were promoting) was either the Business and

Professions Committee (B&PC) in the Senate or the B&PC in the Assembly. When seeking an author we often approached members of these committees as well. The intent was to get early support and move the bill through the legislative process.

5) Finding support. In addition to sponsors, it helps to find other organizations that will support a bill through (1) letters, (2) testimonies, and (3) lobbying efforts. We collected binders of letters from individuals, municipalities, regulators, and corporations. These included Chevron, the Chemical Industry Council of California, and Genentech. The letters provided a lot of anecdotes on how this perverse law affected people. The volume of the letters attracted the attention of legislators.

Once a bill is scheduled for a vote, testimonies at the hearing can be very compelling. The city of San Diego, California Fire Prevention Officers Association, a division of the California Fire Chief Association, along with various individual department fire chiefs and fire marshals, individuals impacted by current law, and others have testified at legislative hearings over the years. The Dean of Engineering at University of California—Davis testified on behalf of the University of California at the hearing in January 2012, stating that the state is educating engineers who cannot work in their own professions. The dean was shocked to find out that the legislators still would not vote to support changing the current law.[2]

The third level of support is when companies/ organizations involve their lobbyists. However, we found it very difficult to convince companies that had lobbyists in Sacramento to get engaged at this higher level.

6) Other curve balls. We experienced numerous unexpected challenges. One such curve ball happened in 1997, when chemical engineers tried independently (without CLCPE) to change the law. The Engineering Board required that we provide a Sunrise Package[3]. Essentially converting engineers from Title Act (defined in by BPELSG Board Rules) to Practice Act (defined in the Professional Engineers Act) is viewed as creating a new engineering discipline. We were given one week to prepare this. Fortunately, CLCPE had prepared something similar 10 to 15 years earlier, which we were able to mine and then create the required Sunrise Package. The Engineering Board still did not support our bill. Note: Keeping a good archive of material produced in supporting legislation could have unexpected value in the future. The key is being able to find the needed archived information when it is needed. Another advantage is that many of the same people remained engaged in CLCPE for long periods.

---

[2] Incidentally, the western region of ASCE sent their lobbyist to testify in opposition at the same hearing where the UCD dean testified.

[3] Sunrise is a process under which an occupation or profession wishing to receive state certification or licensure must propose the components of the legislation, along with cost and benefit estimates of the proposed regulation

What We Proposed and Why

Our approach to changing legislation evolved over the years. As mentioned earlier, chemical engineers made an attempt in 1997 to raise chemical engineering to a Practice Act (defined by statute). Our opposition protested that our attempt was part of a turf battle among the engineering disciplines. Legislators do not want to be drawn into controversies (e.g., infighting). Our next attempt involved CLCPE, with the intent to define all the Title Act disciplines as Practice Act disciplines. Because we had a coalition involving more than 15 engineering societies, we were able to combat this accusation. Still we did not succeed.

We also shifted our approach and proposed legislation that was closer in alignment with generic engineering. Specifically, we proposed that engineers should be allowed to do work based on their education, experience, or testing. In talking to legislators and to our lobbyist, we learned that we needed to approach any change in engineering statute with baby steps. Adopting generic licensure language (similar to other states or similar to Model Law proposed by the National Society of Professional Engineers (NSPE) was too much at one time. Because of the heavy influence by civil engineering companies in the state, we drafted bills that would continue all the advantages afforded to civil engineering. Unfortunately, our opposition would not negotiate.

In 2000, CLCPE had caught the attention of the legislature sufficient enough that they commissioned a study of the two-tier approach (Title Act/Practice Act) used in California. The study considered natural overlap between the professions, whether there was enough distinction between the engineering disciplines to support this unique regulatory structure, and whether the three Practice Act disciplines (civil, mechanical, and electrical engineering) posed more of a risk to the public (relative to the other engineering disciplines) and thus justified the existing two-tier structure. (See **www.CLCPE.org** for study by the Institute for Social Research.)

The study strongly supported changing the existing regulatory structure. The task of making this change fell to the Engineering Board. The bill it sponsored was opposed by American Consulting Engineers Council (ACEC) - California (a contractors' association) and Professional Engineers in California Government (PECG, a public employee union). Both were well-funded with strong ties to legislators — and the bill failed. [PECG is allowed to donate directly to candidates and its 13,000 members individually campaign for legislators. That has a big influence on those legislators who might support our position.]

The study marked a shift in how we approach changing legislation; CLCPE did not sponsor bills after the study. Instead, CLCPE and our individual engineering societies supported a succession of bills that were sponsored by the Engineering Board or by companies and organizations that use engineering services.

The disadvantage of this approach is that we had less influence on making amendments to the language in the bills. So while sponsoring our own bills had not been successful, this other approach allowed us to continue to pursue fixing the Professional Engineers Act.

Another challenge we had was finding companies and organizations that would commit their lobbyists. Many of these companies were very willing to prepare letters of support, but were not ready to involve their lobbyists.

End Results

This paper provides numerous lessons learned from decades of experience in trying to change California law to allow all engineers to practice in their own professions without finding a civil (typically) or other Practice Act engineer to take responsible charge of the work. Despite years of effort, we have not succeeded in changing the Professional Engineers Act. We are debating our next steps. It is our hope that the lessons we learned will be helpful to others who find the need to engage their state legislators.

SIMULATION AND EXPERIMENTAL INVESTIGATION FOR BIODIESEL PRODUCTION USING REACTIVE DISTILLATION

Nada B. Nakkash *Sarah R. Al-Karkhi*
Al-Nahrain University, Baghdad – Iraq

Abstract

The present work concerned with studying the behavior of reactive distillation to produce biodiesel (methyl oleate) by the reaction of methanol and oleic acid using homogeneous catalysts H_2SO_4, experimentally and theoretically.

Experimentally, after checking the experimental conditions of biodiesel production using bench experiment, a lab-scale packed reactive distillation column was constructed which consist of a heat resistance glass distillation column 42 cm packing height and 3.5 cm inside column diameter, packed with glass raschig rings of 10 mm length, 6 mm outside diameter, and 3 mm inside diameter at atmospheric pressure.

The effect of many variables on conversion of oleic acid to biodeisel have been studied such as molar ratio of methanol to oleic acid 4:1, 6:1 and 8:1, amount of catalyst 0.6, 1.2 and 1.8 g sulfuric acid/g oleic acid, reaction time 36, 57 and 75 minutes, and reaction temperature 100°C, 120°C and 130°C. The design of experiments by the Taguchi method was considered for performing the minimum numbers of experiments of 9. The best operating conditions obtained when the reaction takes place in the still are MEOH/OLAC feed molar ratio 8:1, catalyst amount 1.2 g sulfuric acid/g oleic acid, time of reaction 57 min and reaction temperature 130°C, with 93.5485% conversion of oleic acid. The Molar ratio (OLAC/MEOH) was the most influential parameter on the conversion of oleic acid, while the time has less effect by compare to other variables.

The properties of biodiesel (methyl oleate) such as viscosity, flash point, density and carbon residue were measured experimentally and compared with ASTM standard for biodiesel and petrol diesel. The comparison gives that methyl oleate ester could be used as alternative diesel.

Theoretically, an equilibrium model (EQ) was simulated using MATLAB (R2010a) to solve MESHR equations. UNIQUAC liquid phase activity coefficient model is the most appropriate model to describe the non ideality of OLAC-MEOH-MEOL-H_2O system. The results of the developed model were compared with the results of experimental work which give a good agreement according to their linear correlation coefficient r and multiple coefficient of determination R^2 with percentage error of 2.5%.

Keywords: Biodiesel, Esterification, Reactive Distillation, Taguchi Method, equilibrium model

Introduction

Biomass derived diesel fuel, termed biodiesel, can replace petroleum- based diesel fuels. Environmental pollution has raised the concern on the search for the alternative energy sources. Biodiesel has drawn significant attention since it holds various advantages compared with fossil fuel in terms of renewability, non-toxicity, better lubricant and biodegradability characteristic. Also, biodiesel has high cetane number and low sulfur content, hence it will prolong the machine's lifetime [1,2,3]. Biodiesel is considered to be renewable, since the carbon in the oil or fat originated mostly from carbon dioxide in the air, and it is considered to contribute much less to global warming than fossil fuels. Diesel engines operated on biodiesel have lower emissions of carbon monoxide, unburned hydrocarbons, particulate matter, and air toxics than when operated on petroleum-based diesel fuel [4]. The combustion of all types of fuel, including fuels produced from biomass, releases carbon dioxide into the atmosphere, the burning of biomass does not cause any net increase in carbon dioxide concentration. The reason for this is that plants use carbon dioxide from the atmosphere to grow (photosynthesis) and the carbon dioxide formed during combustion is balanced by that absorbed during the annual growth of the plants used as the biomass feedstock. In the case of fossil fuels, the carbon content of these fuels has been fixed and contained in the earth's crust for millions of years. By burning these fossil fuels the formerly harmless carbon buried in the earth's crust is released into the atmosphere as carbon dioxide resulting in a net increase in the carbon concentration that leads to global warming [5].

The higher cost of biodiesel is due to its production mostly from expensive raw materials and consists of complicated process units. Therefore, it is necessary to develop a process in order to produce biodiesel more efficient and economical. To solve this problems, non-edible oils are suitable for biodiesel production, because edible oils are already in demand and too expensive than diesel fuel. Plants contains non edible oil is considered to be the wonder biodiesel feed stock because of rapid in growth, higher seed productivity, it grow in arid, semiarid and wasteland, it requires little water and fertilizer, can survive on infertile soils, and is not browsed by cattle. Oleic acid is used as a raw material to produce biodiesel because oleic acid is widely found in various plants.

Biodiesel has been synthesized by endothermic esterification reaction of oleic acid and methanol in presence of concentrate H_2SO_4 as a catalyst [6].

The esterification of fatty acid and alcohol into fatty acid ester (biodiesel) is usually conducted in a batch reactor. However the production of biodiesel from esterification reaction in the conventional batch reactor has many problems because of its low conversion, heavy capital investments and high energy costs so this process is not economically [7].

Biodiesel is traditionally produced by batch process using well mixed stirred tank and a series of separation equipments; batch process is slow, labor intensive, and costly for a very-large scale process [8,9]. In large scale continuous process are the more efficient processes, which able to reduce post reaction cleaning and total processing time [10]. However, in most existing continuous processes the reaction and product separation occurs separately in conventional configuration which is not economical since it requires high capital and energy cost. Hence, a more effective and efficient process system, development of a novel configuration of biodiesel production which enables the integration of the chemical reaction and product purification in single equipment is necessary. This method is known as a reactive distillation (RD) offers benefits by integrating distillation and reaction in one unit. It reduces separation steps, lower the capital and operating cost, and shift the equilibrium towards the products [11]. Batch reactive distillation (BRD) is a simple experimental tool to quickly evaluate the feasibility of reactive distillation for a reaction of interest.

Maya, A., et. al. [12] Fischer esterification of the mixture of palm fatty acids with isopropanol can be achieved in 80% conversion, under conditions of batch reactive distillation and methanesulphonic acid as catalyst.

Kumar, R. and Mahjani, M.S. [13] evaluated the applicability of batch reactive distillation for esterification of lactic acid with *n*-butanol to synthesize *n*-butyl lactate in the presence of cation exchange resins as a catalyst, an equilibrium stage model is formulated, and simulation results were compared with the experimental results.

Edreder, E.A., et. al. [14] evaluated the performance of batch reactive distillation to produced lactic acid by hydrolysis reaction of methyl lactate. Minimum time optimization problem was developed incorporating a process model within gPROMS software.

Kusmiyati, et. al. [7] determined the best conditions for biodiesel production from the esterification reaction of oleic acid, methanol and sulfuric acid as a catalyst by batch reactive distillation, and the effect of several variables was studied.

Present Work

The experimental part of the present consist of bench experiment to check the experimental conditions for biodiesel production and construct a lab-scale packed reactive distillation column which is used for the production biodiesel by esterification of oleic acid with methanol using sulfuric acid as a catalyst. Different variables such as molar ratio methanol to oleic acid, amount of catalyst, reaction time, and reaction temperature have been studied in order to find the best conditions for biodiesel production by reactive distillation. The design of experiments strategy using Taguchi orthogonal design matrix has been used to minimize the number of experiments from 81 to 9 for 3 level and 4 operating conditions and covers a wide range of operating conditions.

Different methods of analysis of product have been used and the characteristics of biodiesel such as flash point, viscosity, density, and carbon residue have been studied.

The theoretical part emphasized to construct mathematical model based on equilibrium model (EQ) for unsteady state multicomponent packed reactive distillation for biodiesel production using rigorous method. The experimental results are compared with the results of the equilibrium model.

Experimental Work

-Materials

Oleic acid extra pure was procured from "Loba Chemie", Mumbai, India (IV = 92, AV = 200). Anhydrous methanol (MEOH), 99.8%, was procured from "Scharlau Company", Spain, potassium hydroxide (KOH) 0.1 M solution procured from BDH CVS Chemicals, sulfuric acid (H_2SO_4) from "Gainland Chemical Company", Factory Road, Sandycroft Deeside, and Phenolphthalein "The British- Drug House LTD", B.D.H. Laboratory Chemicals Division, Poole, England.

-Chemical Analysis

Each sample of the oil phase was taken from the bottom, at three equal intervals from the time of reaction for analysis by GC and titration. The samples were analyzed using Gas Chromatography (GC). The type of GC was Packard equipped with a capillary column SE-30 (5m length 0.32cm ID) and a flame ionization detector (FID). The temperatures of injector detector were set at 275°C and 350°C respectively. Helium gas (He) of high purity was used as the carrier gas with flow of 30 ml / min. The oven temperature was initially set at 100°C, increasing to 300°C, the temperature increase for both ramps was set at 5°C/min.

The titration is performed by an alkaline 0.1M KOH solution. The volume of KOH solution consumed is reported, and the acidity of the sample is calculated using the equation (1) and the conversion of oleic acid was calculated using the equation (3). Oleic acid (FFA) conversion for both GC and titration were compared, and a good linear regression is obtained (linear Correlation coefficient r of 0.9983, and multiple coefficient of determination R^2 of 0.9964).

$$AV = \frac{mg_{KOH}}{g_{FFA}} \quad (1)$$

$$AV = \frac{V_{KOH} * C_{KOH} * M_{wtKOH}}{1g_{FFA}} \quad (2)$$

$$Conversion\% = \frac{AV_{t=0} - AV_{t=t}}{AV_{t=0}} * 100 \quad (3)$$

-Experiments

Bench experiment to produce biodiesel (methyl oleate) was carried out in a batch reactor, consists of 500ml three-necked round flask one neck handle thermometer and the other handling a total reflux condenser, heat was supplied by a glycerin oil bath. Oleic acid and methanol are introduced to the flask with a molar ratio of 8:1 methanol to oleic acids and mixed continuously by magnetic stirrer with a measured amount of catalyst to produced biodiesel (methyl oleate); sulfuric acid is used as a catalyst with an amount of 1 g sulfuric acid/g oleic acid. The mixture is well mixed and heated until reaches the reaction mixture temperature. After a certain time a sample was taken and analyzed using gas chromatography. The results and the operating conditions are given in Table (1).

Table (1): Results bench experiment

Time (min)	Temperature (°C)	%Conversion
40	68	70.6522
50	100	82.0480

The experiments were carried out using laboratory scale batch reactive distillation column Figures (1). The unit consists of a still pot (D), which is heated using glycerin bath (C) that takes heat from a magnetic stirrer. The distillation column is located above the still pot, packed with raschig rings. At the top of column, a double pipe water-cooler condenser is connected, which is used to condense the vapor leaving the top of column. The total height of packing is 42cm, the inside diameter of column is 3.5 cm, packed with glass raschig rings of 10 mm length, 6 mm outside diameter, and 3 mm inside diameter.

Oleic acid was charged into the still pot (reactor flask), the acid catalyst (H_2SO_4) of 98% purity was added to methanol and the mixture was charged to reactor flask. The reactants (oleic acid, methanol and the catalyst) were mixed in 2L flask, heated in a glycerin bath and kept at reaction temperature and atmospheric pressure. During this period, samples were taken at the specified time to analysis by GC and for titration.

At the end of the distillation process the content in the column is removed in graduated cylinder to separate water phase from the methyl oleate (biodiesel) phase if water is available, where esterification of fatty acid is a reversible reaction and water is formed. Removal of water can drive the reaction equilibrium to the completion and therefore increase the conversion. By removing water as by product the equilibrium is shifted towards ester methyl oleate (biodiesel) formation. The biodiesel will always be separated at the bottom of the reactive distillation column. Water is present as a side product and typically is removed as top product due to its lower boiling point, together with the methanol.

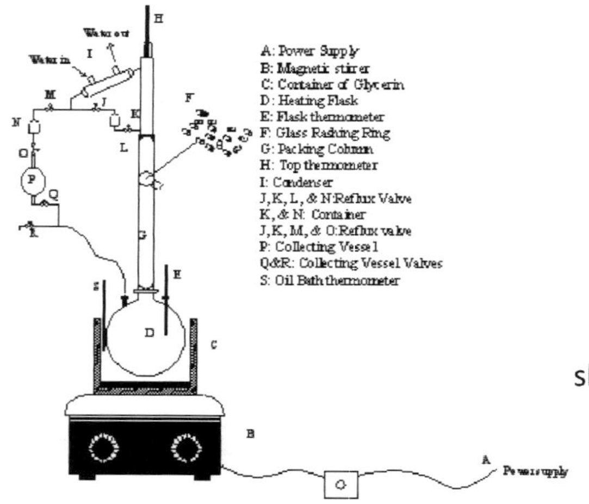

Figure 1. Schematic Sketch of RD Unit

In the present work experimental design by the Taguchi approach was considered with orthogonal array design to show the effects of four variables, including the molar ratio of methanol to oleic acid, amount of catalyst, reaction time and reaction temperature, on the production of biodiesel (methyl oleate).

Four selected variables at three-levels, i.e. L-9 (3^4) experimentally studied as given in Table (2). The diversity of factors was studied, too by crossing the orthogonal array of the control variables Table (3).

The results are statistically analyzed using the signal-to-noise (S/N) ratio to determine the percentage contribution of individual variables to the average oleic acid conversion.

Mathematical Model

The mathematical equilibrium model for batch reactive distillation was formulated using the following assumptions:

1- Constant pressure drop across the column.
2- Hold-up per stage equal to liquid hold up on stage (i.e. vapor phase molar hold-up is neglected).
3- Each stage is considered as a continuous stirred-tank reactor (CSTR).
4- There is heat transfer in the reboiler and in the condenser, but the interior stages of the column are adiabatic.

The schematic model of jth segment is shown in Figure (2).

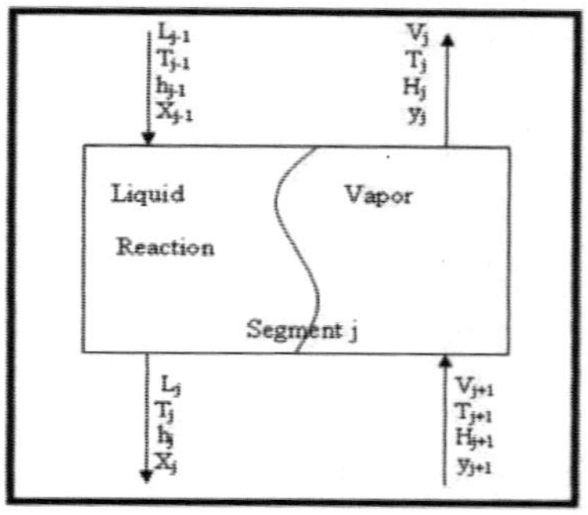

Figure 2. Schematic diagram of equilibrium segment

- *MESHR Equations*

M. Total and component material balances.

$$\frac{dM_j}{dt} = V_{j+1} + L_{j-1} - V_j - L_j + R_j \quad (4)$$

$$\frac{dM_j x_{ij}}{dt} = V_{j+1} y_{i,j+1} + L_{j-1} x_{i,j-1} - V_j y_{i,j} - L_j x_{i,j} + R_{i,j} \quad (5)$$

E. Equilibrium relation

$$y_{i,j} = K_{i,j} x_{i,j} \quad (6)$$

S. Summation equations

$$\sum_{i=1}^{c} x_{i,j} = 1, \sum_{i=1}^{c} y_{i,j} = 1 \quad (7)$$

H. Enthalpy equation

$$\frac{dM_j H_j}{dt} = V_{j+1} H_{j+1} + L_{j-1} h_{j-1} - V_j H_j - L_j h_j + R_j Hr \quad (8)$$

R. Reaction rate equations [15,6].

$$R_{OLAC} = -\frac{d[OLAC]}{dt} = k_1 [_{OLAC}] * W_{cat} \quad (9)$$

Where the kinetic constant k_1 in equation (9) is given by the Arrhenius equation [10]:

$$k_1 = 1.27 \exp\left(\frac{-13300}{RT}\right) \quad (10)$$

The concentration of oleic acid is replaced by activity, equation (9) becomes:

$$R_{OLAC} = k_1 a_{OLAC} * W_{cat} \quad (11)$$

The derivative of the rate reaction is given in Yadav et. al., [16].

-*Vapor-Liquid Equilibrium Relation*

For non-ideal mixture additional variables such as γ_i (activity coefficient) and ϕ_i (fugacity coefficient) appears to represent the degree of deviation from ideality.

$$y_i = \frac{K_i \gamma_i}{\phi_i} x_i \quad (13)$$

The results of Vapor Fugacity Coefficient by Redlich/Kowng and Peng-Robinson cubic equations of state show that the vapor phase has ideal gas behavior and the fugacity coefficient $\cong 1$ [6].

NRTL, UNIQUAC and UNIFAC models have been used to calculate the activity coefficient to select the appropriate liquid phase activity coefficient model for OLAC-MEOH-MEOL-Water System, different activity coefficient models were compared with the experimental results taken from Oliveira et. al., [17]. The experimental data was at atmospheric pressure. The experimental boiling point temperature of the system was compared with the predicted boiling point temperature from each of the activity coefficient models.

Figure 3 shows that the UNIQUAC points nearly fall on the diagonal, indicating that the UNIQUAC liquid phase activity coefficient model is the most appropriate model to describe the non ideality of OLAC-MEOH-MEOL-H_2O system. Parameters of Activity Coefficient Model are available in Table A-1 and A-2.

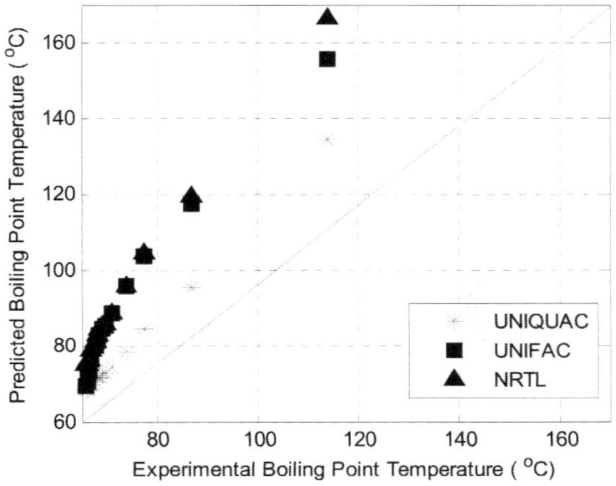

Figure 3. Comparison between experimental and predicted boiling points.

-*Enthalpy Calculation*

The molar enthalpies of the liquid h^L and vapor h^V were calculated using the following equation [18]:

$$h^V = \sum_{i=1}^{n} y_i \int_{T_{ref}}^{T} CP_i^V dT \quad (14)$$

$$h^L = \sum_{i=1}^{n} x_i \left(\int_{T_{ref}}^{T} CP_i^V dT - \lambda_i \right) + H_{mix} \quad (15)$$

-Vapor Pressure Calculation

The vapor pressure of each component for the present system was calculated using Antoine equation.

$$LnP^o = A - \frac{B}{T + C} \qquad (16)$$

Constants of Antoine equation for each component are available in Table A-3.

-Holdup

The equilibrium model was considered for tray columns, to change the concept of the equilibrium stage to packed columns, the idea of the Height Equivalent to a Theoretical Stage (HETP) was considered. HETP value represents a certain bed length of a packing equivalent to one theoretical stage, HETP for the random packing [19].

$$HETP, ft = 1.5 D_P, in \qquad (17)$$

Molar holdups in condenser system and on the column stages based on constant volume holdup G_j :

$$M_j = \frac{G_j}{\sum_{i=1}^{N} \frac{x_i \cdot M_{w_i}}{\rho_i}} \qquad \text{Where j=1 to N-1} \qquad (18)$$

The holdup in reboiler based on the initial charge to the reboiler (M°) and it is given by [13]:

$$M_N = M_N^\circ - \sum_{j=1}^{N} M_j - \int_0^t D_t dt \qquad (19)$$

Stages numbered down from top, consider N=1 for condenser.

Results and Discussion

-Effect of Molar Ratio

Molar ratio of methanol to oleic acid is one of the most important variables that are affecting the conversion of oleic acid. In the present work methyl oleate (biodiesel) was produced from reaction of oleic acid and methanol using various molar ratios of methanol/oleic acid of 4:1, 6:1 and 8:1. Stoichiometrically, the methanol/oleic acid molar ratio required was 1:1, but in practice this was not sufficient to complete the reaction. Higher amount of methanol was required to drive the reaction to completion at faster rate.

Figure (4) shows that as the molar ratio of methanol/oleic acid increases the % average conversion of oleic acid increases, the highest % conversion of oleic acid is 92.4265, for experiment number 8 and the molar ratio of methanol/oleic acid is 8:1. This increase is explained by the shift in the equilibrium which causes by the excess of methanol, so the effect of backward reaction is small, and can be neglected when using excess methanol (higher molar ratio of methanol to oleic acid).

This increase also can be explained by the emulsion system changes from dispersion of methanol into oleic acid towards dispersion of oleic acid into methanol. This transformation results cause an increase in the interfacial area up to a point above which the interfacial area starts to decrease as the cavitations in methanol phase is much easier than in oleic acid phase due to viscosity difference [20].

-Effect of Catalyst Amount

The amount of catalyst has significant effects on the conversion of esterfication reaction. The reaction rate of esterfication reaction is directly proportional to the amount of a catalyst, so the catalyst is used to enhance the reaction rate and conversion. It gives lower activation energy, thus, more products will be formed, and the amount of sulfuric acid employed as a catalyst is related to formation of H^+ that catalyzes the reaction [21]. Increasing the amount of catalyst increases the reaction rate and consequently reduces the time to achieve a high conversion.

Different amounts of catalyst of 0.6, 1.2 and 1.8 g sulfuric acid/g oleic acid have been used. Figure (5) shows the effective amount of catalyst is 1.2, when the amount of catalyst is 0.6 the % average conversion of oleic acid is 80.0845%, increasing the amount of catalyst to 1.2 the % average conversion of oleic acid becomes 84.3509%. For further increase in the amount of catalyst above 1.2, there is no significant increase in the % average conversion of oleic acid, thus increasing the amount of catalyst to 1.8 the % average conversion of oleic acid becomes 84.105%.

The highest % conversion of oleic acid is 92.4265% for experiment number 8 using catalyst amount of 1.2 g sulfuric acid/g oleic acid.

- Effect of Reaction Time

Different reaction times of 36, 57 and 75 minutes have been studied. Figure (6) shows that when the reaction time increases the % average conversion of oleic acid increases up to 57min and decreases for further increase in time.

Thus, the % average conversion of oleic acid increases from 81.3935% to 84.1518% with increasing the time of reaction from 36 to 57 min but decreases to 82.9953% when the time increased to 75 min. The explanation of this observation is due to the loss of methanol from the mixture during the reaction.

The highest %conversion of oleic acid is 90.4630% for experiment number 9 and reaction time 57min.

-Effect of Reaction Temperature

Reaction temperature is another critical variable, in the present work the column operates at atmospheric pressure with different reaction temperatures of 100°C, 120°C and 130°C. Figure (7) shows that the reaction is endothermic and when the temperature increases the % average conversion of oleic acid increases too.

The highest % conversion of oleic acid is 92.4265% for experiment number 8 and reaction temperature 130°C. This is explained as the reaction rate increases a behavior of reactions with higher activation energy are favored by higher temperatures. When operating the column at temperature higher than 130°C, the product will degrade and by lowering the temperature a better color of product is obtained.

-Percentage Contribution of Individual Variables

In Taguchi method, the signal- to-noise (S/N) ratio is used to measure the quality characteristics deviating from the desired value using the terms of 'larger-the-better' [6].

The contribution of an experimental variable was calculated from the maximum difference in the values between the mean S/N ratios at each level Table (4).

The order of influence of the parameters in terms of the conversions was:

Molar ratio (MEOH/OLAC) > Reaction Temperature (D) > Catalyst amount (g sulfuric acid/g oleic acid) (B) >Time (min) (C), Figure (8) illustrates the percentage contribution of individual variables on variation in oleic acid conversion.

A larger mean S/N ratio indicates a greater effect of the control variable at that level on the conversions of oleic acid. Therefore, the best conditions for the largest conversions of oleic acid were A3, B2, C2 and D3. In other words, based on the S/N ratio, the best variables were A Molar ratio (MEOH/OLAC) at level 3 (1:8), B (Catalyst Amount) at level 2 (1.2 g sulfuric acid/g oleic acid), C (Time) at level 2 (57 min) and D (reaction temperature) at level 3 (130°C).

Comparison of Biodiesel with ASTM and Petroldiesel Fuels

The best physical properties of biodiesel obtained from the present work at molar ratio of methanol to oleic acid 1:8, catalyst amount 1.2 g sulfuric acid/g oleic acid, time 57 min and reaction temperature 130°C are compared with ASTM and Petrol diesel Fuels properties, Table (5).

Comparison of Experimental and Equilibrium Model Results

For the different variables studied the results of experimental and theoretical equilibrium model were compared. Figure (9) shows the composition profile of oleic acid and methyl oleate (biodiesel) with time in still and Figure 10) shows the % conversion of oleic acid with time for the best conditions of molar ratio 8:1, catalyst amounts 1.8, reaction time 57min and reaction temperature 130°C.

The best of fit of the experimental results to theoretical equilibrium model was assessed by comparing the experimental conversion of oleic acid with the theoretical equilibrium model conversion of oleic acid, also good linear regression between experimental and theoretical results according to linear correlation coefficient r and multiple coefficient of determination R^2 for the best operating conditions are 0.9697 and 0.9381 respectively, with percentage error of 2.5%.

Conclusions

The average conversion of oleic acid increases with increasing methanol to oleic acid feed molar ratio, amount of catalyst, time of reaction, and the reaction temperature.

For all the variables studied and according to Tanguchi method the best parameters conditions are methanol to oleic acid feed molar ratio 8:1, catalyst amount 1.2 g sulfuric acid/g oleic acid, time of reaction 57 min and reaction temperature 130°C, with oleic acid conversion of 93.5485%. Also, according to Tanguchi method the molar ratio methanol to oleic acid was the most influential variable on the average conversion of oleic acid and the time of reaction has a less effect by compare with other variables.

The reaction shows the typical behavior of reactions with high activation energy that are favored by higher temperatures (endothermic reaction).

Properties of the biodiesel obtained from the present work such as viscosity, flash point, density, and carbon residue show that the biodiesel formed can be used as a fuel.

The UNIQUAC liquid phase activity coefficient model is the most appropriate model to describe the non ideality of OLAC-MEOH-MEOL-H_2O system.

There is a good agreement between the experimental and the developed equilibrium model results in composition profile of oleic acid and methyl oleate (biodiesel) and in % conversion of oleic acid with error of 2.5%.

References

1. Budiman, A., Kusumaningtyas, R.D., Sutijan; Rochmadi and Purwono, S. (2009) "Second Generation of Biodiesel Production from Indonesian Jatropha Oil by Continuous Reactive Distillation Process", AJCHE, vol.9, No. 2, 35-48.

2. Boucher, M.B., Unker, S.A., Hawley, K.R., Wilhite, B.A., Stuart, J.D. and Parnas, R.S. (2008) "Variables affecting homogeneous acid catalyst recoverability and reuse after esterification of concentrated omega-9 polyunsaturated fatty acids in vegetable oil triglycerides", Green Chem., 10, 1331–1336.

3. Kiss, A.A., Dimian, A.C. and Rothenberg, G. (2008) "Biodiesel by Catalytic Reactive Distillation Powered by Metal Oxides", Energy & Fuels, 22, 598-604.

4. Gerpen, J. (2005) "Biodiesel processing and production", Fuel Processing Technology 86, 1097– 1107.

5. Uriarte, J.F. (2010) "Biofuels from Plant Oils", the ASEAN Foundation, Jakarta, Indonesia.

6. Al-Karkhi, S.R. (2012) "Simulation and Experimental Investigation for Production Biodiesel using Batch Reactive Distillation", M. Sc. Thesis, Al-Nahrain University, Baghdad, Iraq.

7. Kusmiyati, Sugiharto, A. (2010) "Production of Biodiesel from Oleic Acid and Methanol by Reactive Distillation", Bulletin of Chemical Reaction Engineering & Catalysis (BCREC).

8. Kiss, A.A., Omota, F., Dimian, A.C. and Rothenberg, G. (2006) "The heterogeneous advantage: biodiesel by catalytic reactive distillation", Springer, 40, 141-150.

9. Singh, A. P., Thompson, J. C., He, B. B. (2004) "A Continuous-flow Reactive Distillation for Biodiesel Preparation from Seed Oil", ASAE/CSAE Meeting presentation, 1-4 August.

10. Bisowarno, B.H., Tian, Y.C., and Tade, M.O. (2004) "Application of Side Reactors on ETBE Reactive Distillation", Chem. Eng. J., 99, 35–43.

11. Kusumaningtyas, R.D., Budiman, A., Rochmadi, Sutijan, Purwono, S., "Experimental Investigation of Biodiesel Synthesis from palm Oil using Reactive Distillation Process", Regional Symposium on Chemical Engineering, RSCE, Manila, Philippine December (2009).

12. Maya, A., Viales, L.C., Arias, M.M., Mata, C.J.F. and Segreda (2006) "Kinetics of the Fischer esterification of palm fatty acids with isopropanol and butyl- cellosolve", Ciencia y Technology, 24(2): 175-181.

13. Kumar, R. and Mahjani, M.S. (2007) "Esterification of Lactic Acid with n-Butanol by Reactive Distillation", Ind. Eng. Chem. Res., 46, 6873-6882.

14. Edreder, E.A., Mujtaba, I.M. and Emtir, M. (2010) "Optimization of Batch Reactive Distillation Process: Production of Lactic Acid, 20th European Symposium on Computer Aided Process Engineering-ESCAPE20.

15. Sendzikiene, E., Makareviciene, V., Janulis, P., Kitrys, S. (2004) "Kinetics of free fatty acids esterification with methanol in the production of biodiesel fuel", Eur. J. Lipid Sci. Technol. 106, 831–836.

16. Yadav, P.K.S. Singh, O., Singh, R. P. (2010) "Palm Fatty Acid Biodiesel: Process Optimization & Study of Reaction Kinetics", Journal of Oil Science, 59,11,575-580.

17. Oliveira, M.B., Miguel, S.I., Queimada, A.J., and Coutinho, J.A.P. (2010) "Phase Equilibria of Ester + Alcohol Systems and Their Description with the Cubic-Plus-Association Equation of State", Ind. Eng. Chem. Res., 49, 3452-3458.

18. Walas, S.M. (1985) "Phase equilibria in chemical engineering", Butterworth Publications, London.

19. Seader, J.D., and Henley, E.J. (1998) "Separation process principles", John Wiley and Sons, Inc., New York.

20. Mahamuni, N.N. and Adewuyi, Y.G. (2010) "Application of Taguchi Method to Investigate the Effects of Process Parameters on the Transesterification of Soybean Oil Using High Frequency Ultrasound", Energy Fuels, 24, 2120-2126.

21. Lucena, I.L., Silva, G.F. and Fernandes, F.A.N. (2008) "Biodiesel Production by Esterification of Oleic Acid with Methanol Using a Water Adsorption Apparatus", Ind. Eng. Res., 47, 6885-6889.

Symbols	**Notation**
AV	Acid value.
CP^L	Specific heat of liquid
CP^V	Specific heat of vapor
D_p	Out side diameter of packing
h^L	Total enthalpy of liquid phase
h^V	Total enthalpy of vapor phase
h_L	Liquid hold up in Packing.
H_{mix}	heat of mixing
HETP	Height equivalent to theoretical plates.
L	Liquid flow rate
M_{cat}	Mass of catalyst
M_{wi}	Molecular weight
N_T	Number of stages
N_c	Number of components
P	Pressure
P°	Vapor presure
Q	Heat duty
R	Gas constant = 8.314
R_{FFA}	Reaction rate equation
R^2	Coefficient of Multiple determination
r	Linear correlation coefficient for sample
r_i	Volume parameter of component i in UNIQUAC and UNIFAC models
T	Temperature
T_{ref}	Reference temperature
T	flow time
V_{KOH}	Volume of KOH solution consumed from titration
V	Titration volume.
V	Vapor flow rate
W_{cat}	Weight of sulfuric acid
x_i	Liquid mole fraction
y_i	Vapor mol fraction

Greek Letters

ν	kinematic viscosity at 40°C
ϕ_i	Fugacity coefficient of component i in mixture
γ_i	Activity coefficient of component i in mixture
ρ	Liquid molar density

Subscripts

i, j	Component indices
L	Liquid
V	Vapor

Superscripts

Sat	Saturated
L	Liquid
V	Vapor

Table (2): Design experiments, with four parameters at-three level, for the production of methyl oleate (Biodiesel)

variables	Levels		
	1	2	3
A Molar ratio (OLAC/MEOH)	1:4	1:6	1:8
B Catalyst Amount (g sulfuric acid/g oleic acid)	0.6	1.2	1.8
C Time (min)	36	57	75
D Reaction Temperature (°C)	100	120	130

Table (3): Orthogonal array used to design experiments with four parameters at three-levels, L-9(3^4)

Exp.No.	variables and their level			
	Molar ratio (OLAC/MEOH) A	Catalyst Amount B	Time (min) C	Reaction Temperature D
1	1	1	1	1
2	1	2	2	2
3	1	3	3	3
4	2	1	2	3
5	2	2	3	1
6	2	3	1	2
7	3	1	3	2
8	3	2	1	3
9	3	3	2	1

Table (4): The distribution of the four influential variables

Variables	Max. Difference	Contribution (%)
A Molar ratio (MEOH/OLAC)	1.2797	49.3445
B Catalyst Amount	0.4737	18.2656
C Time (min)	0.3355	12.9367
D Reaction Temperature	0.5045	19.4532
Total	2.5934	100

Table (5): Comparison of different biodiesel and diesel

Fuel Properties	Present work	ASTM	Diesel Oil	Method
Flash point, °C (close cup)	165	≥ 130	52-66	ASTM D 93
Kinetic Viscosity, mm2/s, (at 40 °C)	4.457	1.9-4.1	1.9-4.1	ASTM D445
Density g/ml	0.875	0.87-0.89	0.82	ASTM D 1298
Carbon residue. wt%	0.0392	Max. 0.05	-	ASTM D 189
Ester Content, wt%	95.516	-	-	(Felizardo et. al,. 2006)

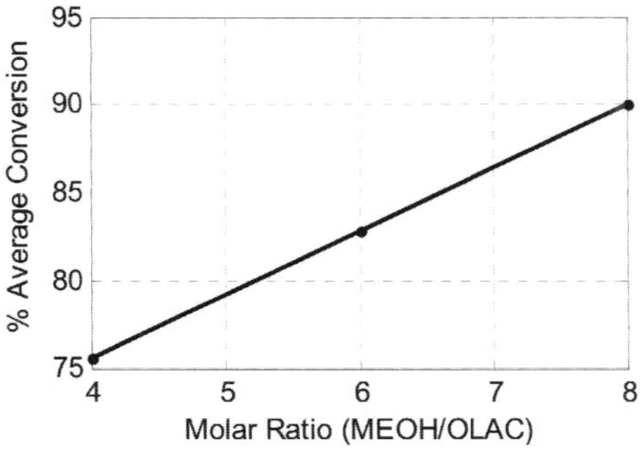

Figure 4. Effect of molar ratio on % average conversion

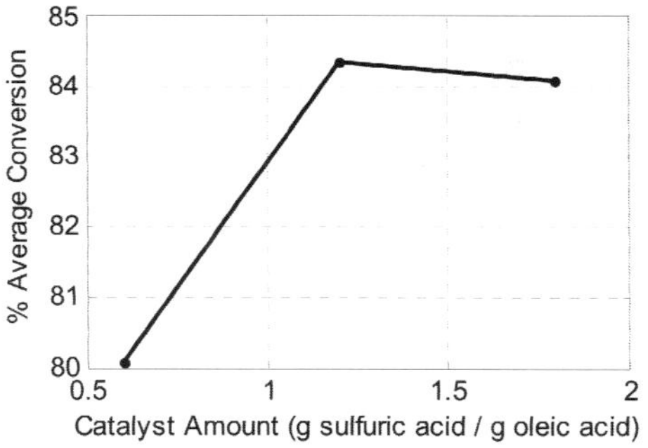

Figure 5. Effect of catalyst amount on the conversion

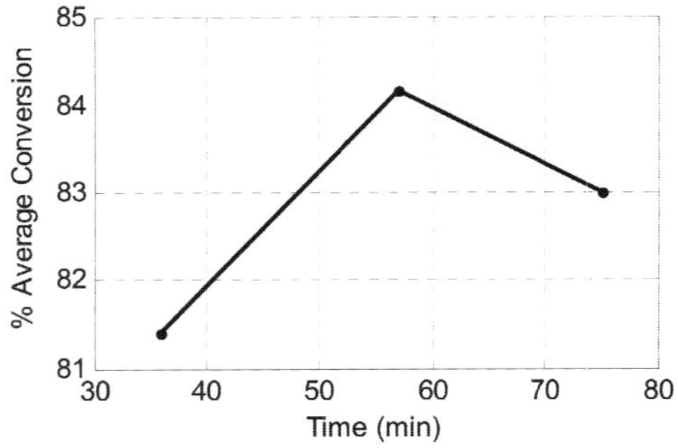

Figure 6. Effect of time on % average conversion

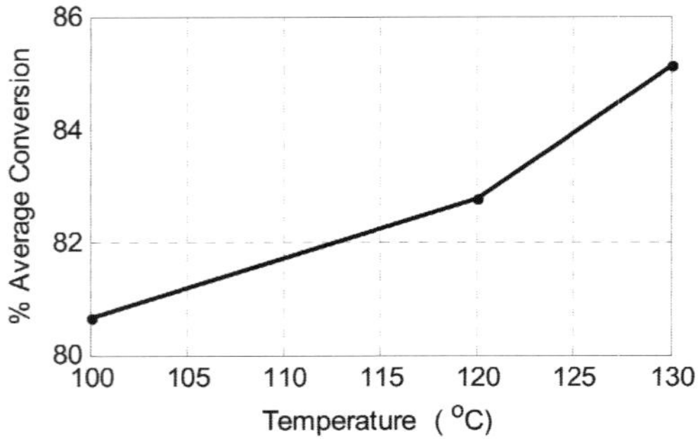

Figure 7. Effect of reaction temperature on % average conversion

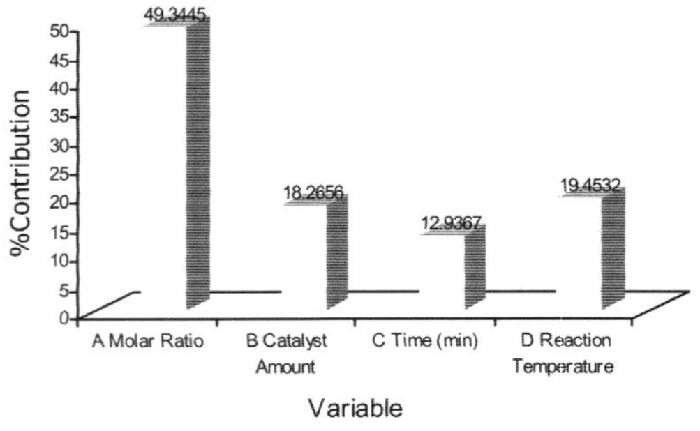

Figure 8. Percentage contribution of individual variables on variation in oleic acid conversion

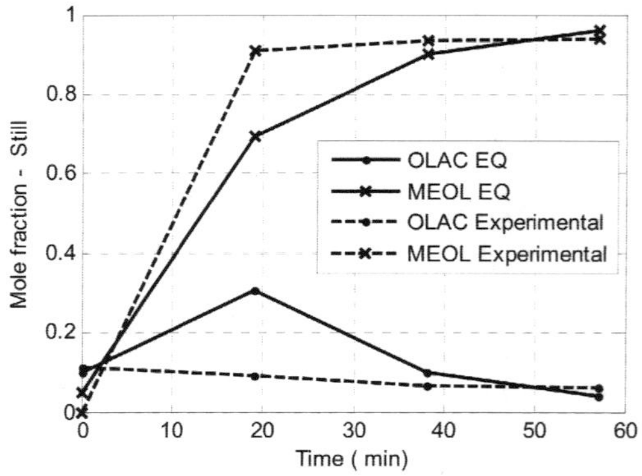

Figure 9. Experimental and theoretical equilibrium model results for composition Profile in the still, molar ratio 8:1, catalyst amount 1.2, 57 min, 130°C

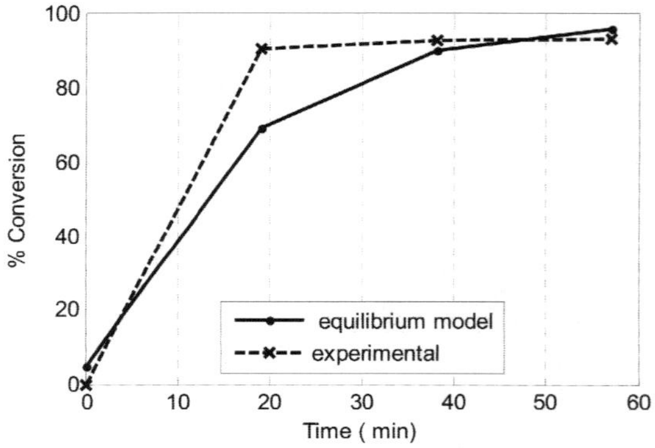

Figure 10. % Conversion profile for experimental and theoretical equilibrium model Best Experiment

Table A.1 NRTL parameters for the binary pairs of components in the reactive mixtures

$i-j$	B_{ij}	B_{ji}	α_{ij}
OLAC- MEOH	199.884	479.688	1.1431
MEOH- H_2O	-24.4933	307.166	0.3001
OLAC- MEOL	37.63835	36.76161	0.2907206
OLAC- H_2O	-44.8289	2497.61	0.2250879
MEOH-MEOL	1388.564	-240.4565	0.399494
MEOL- H_2O	106.4762	2499.963	0.200312

Table A.2 UNIQUAC parameters for the oleic acid – methanol – methyl oleate – water mixture, cal/gmol.

$i-j$	$u_{ij}-u_{jj}$	$u_{ji}-u_{ii}$
OLAC- MEOH	952.028	-149.181
MEOH - H_2O	95.259	-10.377
OLAC- MEOL	154.7875	-133.418
OLAC - H_2O	1123.794	403.7021
MEOH- MEOL	-54.20368	1205.077
MEOL - H_2O	1573.999	481.5153

Table A-3 Vapor pressure constants

Component	Antonio Coefficient		
	A	B	C
OIAC	23.1373	5884.49	-127.26
MEOH	23.4803	3626.55	-34.29
MEOL	22.8313	5948.17743	-96.15
Water	23.1964	3816.44	-46.13

(83b) Molecular Origin of Cellulose Dissolution in Novel Solvent Mixtures

Ritankar Das, Chemical and Biomolecular Engineering, UC Berkeley, Berkeley, CA and Jhih-Wei Chu, Chemical Engineering, University of California, Berkeley, Berkeley, CA

Biomass contains abundant amounts of cellulose as crystalline microfibrils. A limiting step to using cellulose as an alternative energy source, however, is the hydrolysis of the biomass and subsequent transformation into fuels. Cellulose is insoluble in most solvents including organic solvents and water, but it is soluble in some ionic liquids like BMIM-Cl. This project aims to find alternative solvents that are less expensive and are more environmentally benign than the ionic liquids.

Figures: (Above) Setup of glucan chains in randomized solvent mixture. Orange-pyridine, Blue-propanol, Multicolor- glucan chains. (Graph Below) Contacts after equilibration in a Pyridine-Propanol solvent mixture. (Table Below) Contacts after equilibration in a pure solvents.

Pure Solvents	
Solvent	***Contacts***
Propanal	***1632***
Propanol	***1025***
Water	***1106***
Imidazole	***179***
Pyridine	***245***

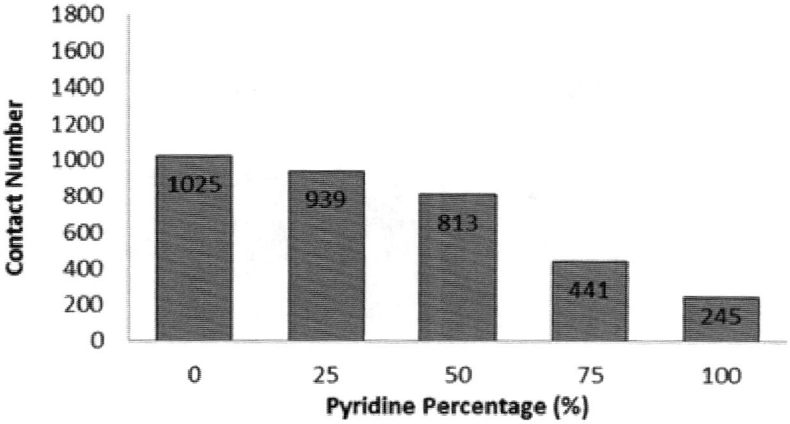

All-atom molecular dynamics simulations were performed on dissociated glucan chains separated by multiple (4-5) solvation shells, in the presence of several novel solvents and solvent mixtures. The solubility of the chains in each solvent was indicated by contacts calculations after the equilibration of the molecular dynamics. It was discovered that pyridine and imidazole acted as the best solvents because their aromatic electronic structure was able to effectively disrupt the inter-sheet interactions among the glucan chains in the axial direction, and because perturbation of the solvent interactions in the presence of glucan chains was minimal.

(83c) Biochemical Engineering Characterisation of a Shaken Micro-Photobioreactor Platform for High-Throughput Development of Microalgae Cultivation Processes

Ebenezer Ojo, Hadiza Auta, Kane Miller, Frank Baganz and Gary J. Lye, Biochemical Engineering, University College London, London, United Kingdom

Micro-scale technologies for early prediction of key bioprocess parameters and scale-up have previously been applied to biopharmaceutical production and production of human cells for therapy. However, their application to sustainable, bio-renewable energy bioprocess development is only now being considered. Here we describe a shaken micro photo-bioreactor platform that provides the required biochemical engineering conditions suitable for optimisation of microalgae cultivation conditions. Cells are cultured in up to 24 parallel wells with working volumes in the range of 3-4 mL. Illumination was provided via a shaking platform equipped with high powered LEDs with adjustable light intensity and the ability to simulate day-night cycles.

Characterization of a biochemical engineering environment in the shaken photoincubator shows 70 % relative humidity as appropriate conditions for evaporation and condensation control. Light intensity variation across the platforms was in the range of $\pm$ 20 $\mu mol.m^{-2}s^{-1}$., with CO_2 gas transfer rate found to be approximately 33.66 $\pm$ 3.0 h^{-1} in the photo-incubator.

However, impact of the engineering conditions (160$\mu mol.m^{-2}s^{-1}$ and 300 rpm shaking speed) showed biomass concentration of 5.87$g.L^{-1}$ for ***Chlorella sorokiniana*** with high fatty methyl esters acid required for bioenergy production. In the absence of light, microbial fermentation using, ***E. coli (hydrocarbon producing)*** was found to be 3.29g/L dry cell weight over cultivation time. In general, the results match with reported literature data.

Adoption of the new micro-scale shaken photobioeactor platform will enable rapid and parallel evaluation of multiple strains of microalgae for their green fuel production capabilities. Along with studies for optimisation of culture conditions this will help reduce bioprocess development costs and timescales.

(83d) Highly Active, Stable, and Recyclable Magnetic Nano-Size Solid Acid Catalysts for Efficient Esterification of Free Fatty Acid in Grease to Produce Biodiesel

Zillillah Zillillah[1], Guowei Tan[1,2] and Zhi Li[1,2], (1)Department of Chemical and Biomolecular Engineering, National University of Singapore, Singapore, Singapore, (2)Singapore-MIT Alliance, Singapore, Singapore

The increase in worldwide energy and fuel demand coupled with the rapid depletion of non-renewable energy such as crude oil have led to a growing interest in biodiesel as a renewable and cleaner alternative to petroleum-based diesel. Biodiesel commonly comprises of fatty acid methyl esters (FAME) produced from vegetable oils ***via*** base-catalyzed transesterification with methanol. Due to the high cost of vegetable oils, the current price of biodiesel is still too high to replace the traditional petroleum-based diesel. Recently, grease has become an attractive feedstock for biodiesel production, since it is a waste at low price and a non-edible resource. However, there is technical challenge that arises when using grease to produce biodiesel. The conventional base-catalyzed transesterification method cannot work well for the grease due to the high free fatty acid (FFA) content (15-40 wt%) present in grease, which often results in soap formation. To solve this problem, grease may be pretreated ***via***an acid-catalyzed esterification to transform FFA into FAME, before performing base-catalyzed transesterification to convert the remaining triglycerides into FAME.

Homogeneous acid catalysts such as sulfuric acid are known for the esterification of FFA but suffer from many drawbacks such as corrosion, pollution, complex downstream neutralization, difficult separation, and no reusability of the catalysts. Heterogeneous solid acid catalysts are thus preferred as greener and recyclable catalysts. However, the reported heterogeneous acid catalysts have been thus far unsatisfactory for the esterification of FFA. Their low activity is mainly caused by the diffusional limitation of the porous materials or the low acid loading and mass transfer limitation of the large size carriers. In this regard, the use of magnetic nanoparticles (MNPs) as catalyst carriers could solve the problems. The high surface-area-to-volume ratio of MNPs could allow for high acid loading and the nano-size of the catalyst could enable good dispersibility and high mass transfer efficiency. Moreover, the magnetic properties could also allow for the convenient and efficient recovery of nanocatalysts under a magnetic field, thus facilitating catalyst recycling.

In this study, active, stable, and recyclable magnetic nano-size solid acid catalysts have been prepared, characterized, and examined, for the first time, as efficient catalysts for the esterification of FFA in grease with methanol to produce biodiesel. Novel acid catalyst consisting of a core of iron oxide MNPs, a poly(glycidyl methacrylate) (PGMA) shell, and sulfonic acid surface functions was facilely fabricated in 98% yield from the corresponding PGMA–MNPs containing epoxide groups on the surface by gentle sulfonation with Na_2SO_3. The catalyst possesses a mean size of 90 nm, high acid capacity of 2.3 mmol H^+ g^{-1}, and excellent superparamagnetic properties. Esterification of FFA (16 wt%) in grease with methanol using this catalyst (4 wt%) gave 96% conversion of FFA to FAME within 2 h. The catalyst was easily separable from the reaction mixture under a magnetic field and showed high stability and recyclability, with no loss of productivity during 10 cycles of esterification. In addition, its catalytic performance and recyclability were also compared with those of other newly fabricated magnetic nano- or micro-size solid acid catalysts and ion-exchange resin with a larger size of support, Amberlyst 15. It was found that the catalyst performance was much better than those of micro-size solid acid catalyst and Amberlyst 15, demonstrating the advantage of the small size of catalyst particles in reducing mass-transfer limitation and providing better accessibility of the catalyst sites for the substrates. The catalyst recyclability was also better compared to those of other magnetic nano-size solid acid catalysts due to the virtue of the stable PGMA shell and the mild sulfonation conditions which give higher stability and better magnetic properties.

(83e) Development of A Hybrid Neural System for Monitoring the Bioethanol Production with Yeast Recycling

William Herrera and Rubens Maciel, Department of chemical processes, Unicamp, Campinas, Brazil

Researches are being conducted worldwide for the development of new technologies for producing liquid fuels from renewable sources. In this context, Brazil, as a bioethanol producer from sugar cane aims, with this alternative, to reduce the crude petroleum consumption as well as the environment pollution. Within the proposed scenario, the goal is to contribute with proceedings and a computer operation technique that will help reducing production costs of ethanol biofuel

In this present work a software sensor system was developed to obtain the biomass substrate and product concentration rates from secondary measurements as: pH, turbidity, CO_2 rate and temperature. The sensor software uses a neural hybrid model to combine a multi;layer percepetion artificial neural network and the mass balance that describes the fermentation kinetic process. Experimental data used for training the hybrid model were obtained from fermentation performed at 30-38 º C, with yeast already used in other fermentation process (recycle cells), with a step of 2 ° C. Fermentation raw material is a is a mixture of hydrolyzed bagasse and cane molasses 75% and 25% volume, respectively. This type of composition is typical of the second generation ethanol production process coupled with the first generation process. The integrated structure allows making the monitoring of bioethanol production in real time.

(83f) Pellets From Lignocellulosic Materials: Briquetting Process Evaluation

Yuri Gonzalez[1], Sahra Rincon[1] and Gabriel Camargo[2], (1)Environmental Engineering, Universidad Libre, Bogota, Colombia, (2)Ingenieria Mecanica, Universidad Libre, Bogota, Colombia

This study looking for the implementation a process to obtain pellet from lignocellulosic materials (Oil Palm waste). These pellets are used as combustion material. The first step in this research was the selection of biomass (Colombia is a main producer in the world of Oil Palm). The waste material was ground, separated by sizing and characterized. The main purpose of this investigation was the briquetting process to give a different use and related with sustainable energy.

An experimental design was elaborated by found the relation between different variables selected (particle size, pressure, and binder). Two particle sizes, two pressure levels and two different substances were employed as binders, like pitch and tar. The pellets was characterized by heating power, density, area bet, pore diameter, pore distribution and physical properties such as strength tensile, among other.

The combustion process was study changed the air flow and followed the ashes content and orsat analysis.

(87a) A DFT Study of NaMgH3 (001) Surfaces : Lattice Dynamics

Fernando A. Soto, Louisiana Tech University, Ruston, LA and Daniela S. Mainardi, Chemical engineering, Louisiana Tech University, Ruston, LA

A DFT Study of $NaMgH_3$ (001) Surfaces : Lattice Dynamics

Authors: Fernando Soto[1], Daniela S. Mainardi[2]

[1]Institute for Micro-manufacturing, Louisiana Tech University

[2]Institute for Micro-manufacturing, Chemical Engineering, Louisiana Tech University

Complex metal hydrides have emerged as one of the most promising solid-state hydrogen storage materials[1] due to their high hydrogen volumetric and gravimetric storage capacities. Mg-based alloys such as $NaMgH_3$ are especially interesting because of their light weight[2] and hydrogen mobility[3]. However, a solid state material must meet a set of technical requirements regarding gravimetric and volumetric densities, operating temperature, delivery pressure, fueling time, cycle life, cost and safety that no system currently satisfies[4]. For this reason, efforts have been devoted to decrease the desorption temperature and enhance the kinetics of complex metal hydrides[5]. For instance, thermodynamics and kinetics have been improved by doping $NaAlH_4$ with transition-metals[6] [7]; however, the improvement in dehydrating and hydrating kinetics occurs at the expense of hydrogen content[8]. Thus, the search for optimal storage materials, catalytic additives and destabilization options remains open.

In spite of considerable experimental and computational work[9-17], a complete understanding of the vibrational properties such as phonons and dispersion of the $NaMgH_3$ (001) surfaces is still lacking. The study of surface phonons provides valuable insight into the surface structure and other phenomena specific to the surface, which can differ from the bulk. A density functional theory study addressing these issues is important in order to achieve a deeper understanding of the relevant mechanisms that could facilitate hydrogen desorption in this material.

[1] Orimo SI, Nakamori Y, Eliseo JR, Züttel A, Jensen CM. Complex hydrides for hydrogen storage. Chemical Reviews. 2007;107:4111-32.

[2] Züttel A. Hydrogen storage methods. Naturwissenschaften. 2004;91:157-72.

[3] Wu H, Zhou W, Udovic TJ, Rush JJ, Yildirim T. Crystal chemistry of perovskite-type hydride NaMgH 3: Implications for hydrogen storage. Chemistry of Materials. 2008;20:2335-42.

[4] Jena P. Materials for hydrogen storage: Past, present, and future. Journal of Physical Chemistry Letters. 2011;2:206-11.

[5] Sakintuna B, Lamari-Darkrim F, Hirscher M. Metal hydride materials for solid hydrogen storage: A review. International Journal of Hydrogen Energy. 2007;32:1121-40.

[6] Bogdanović B, Schwickardi M. Ti-doped alkali metal aluminium hydrides as potential novel reversible hydrogen storage materials. Journal of Alloys and Compounds. 1997;253-254:1-9.

[7] Bogdanović B, Brand RA, Marjanović A, Schwickardi M, Tölle J. Metal-doped sodium aluminium hydrides

as potential new hydrogen storage materials. Journal of Alloys and Compounds. 2000;302:36-58.

[8] Sandrock G, Gross K, Thomas G. Effect of Ti-catalyst content on the reversible hydrogen storage properties of the sodium alanates. Journal of Alloys and Compounds. 2002;339:299-308.

[9] Bouamrane A, De Brauer C, Soulié JP, Létoffé JM, Bastide JP. Standard enthalpies of formation of sodium-magnesium hydride and hydridofluorides NaMgH3, NaMgH2F and NaMgF2H. Thermochimica Acta. 1999;326:37-41.

[10] Bouamrane A, Laval JP, Soulie JP, Bastide JP. Structural characterization of NaMgH2F and NaMgH3. Materials Research Bulletin. 2000;35:545-9.

[11] Bouhadda Y, Boudouma Y, Fennineche NE, Bentabet A. Ab initio calculations study of the electronic, optical and thermodynamic properties of NaMgH3, for hydrogen storage. Journal of Physics and Chemistry of Solids. 2010;71:1264-8.

[12] Hao S, Sholl DS. Hydrogen diffusion in MgH2 and NaMgH3 via concerted motions of charged defects. Applied Physics Letters. 2008;93.

[13] Hao S, Sholl DS. Role of Schottky Defects in Hydrogen and Metal Diffusion in NaH, MgH2, and NaMgH3. The Journal of Physical Chemistry Letters. 2010;1:2968-73.

[14] Ikeda K, Kogure Y, Nakamori Y, Orimo S. Reversible hydriding and dehydriding reactions of perovskite-type hydride NaMgH3. Scripta Materialia. 2005;53:319-22.

[15] Klaveness A, Swang O, Fjellvag H. Formation enthalpies of NaMgH3 and KMgH3: A computational study. Europhysics Letters. 2006;76:285-90.

[16] Shanes DT, Corey RL, Bowman Jr RC, Zidan R, Stowe AC, Hwang SJ, et al. NMR studies of the hydrogen storage compound NaMgH3. Journal of Physical Chemistry C. 2009;113:18414-9.

[17] Wu H, Zhou W, Udovic TJ, Rush JJ, Yildirim T. Crystal Chemistry of Perovskite-Type Hydride NaMgH3: Implications for Hydrogen Storage. Chemistry of Materials. 2008;20:2335-42.

(87b) Production of Hydrogen From Carbon Compounds by Molten Salt Gasification

Lyman Frost[1], Bruce Wilding[2], Neil Camarta[2] and Terry Turner[2], (1)Western Hydrogen Ltd, Calgary, AB, Canada, (2)Western Hydrogen, Calgary, AB, Canada

Western Hydrogen LTD (WHL) is developing a new method for gasification that uses inputs of a carbon bearing compound and water to generate either hydrogen and carbon dioxide or synthesis gas (i.e. hydrogen and carbon monoxide). The research has been done at the Idaho National Laboratory, a US Department of Energy laboratory, under a Cooperative Research and Development Agreement. The gasification process occurs at a pressure of about 2000 psi and a temperature of about 900 degrees centigrade. WHL is currently (November 2012) installing a pilot plant near Fort Saskatchewan, Alberta that will produce 200 MSCF per day of hydrogen at a pressure of 2000 psi from inputs of refinery residuals and water.

The WHL technology uses a sodium salt mixture that is continually regenerated from the input carbon containing compound. The carbon is oxidized using the oxygen from the water input. The hydrogen is generated from the water when the oxygen and carbon unite and from the reaction of the sodium compounds with the water. The carbon dioxide exits the system as a separate stream at elevated pressure and is ready for sequestration or to be used in enhanced oil recovery.

This presentation will review the development of the basic technology and provide a review of the operation of the pilot facility. The information provided will include the data generated at the pilot facility, a mass – energy balance for the pilot facility, and projections for the economics for a proposed commercial demonstration plant.

(87c) Low Carbon Hybrid Energy System: A Possible Solution to China's Future Low Carbon Economy

Zhiyong Tang, Low Carbon Energy Conversion Technology Center, Shanghai Advanced Research Institute, Chinese Academy of Sciences, Shanghai, China, Yuhan Sun, State Key Laboratory of Coal Conversion, Institute of Coal Chemistry, Chinese Academy of Sciences, Taiyuan, China and Tiejun Zhao, Low Carbon Energy Conversion Technology Center, Shanghai Advanced Research Institute, Chinese Academy of Sciences

China's coal dominated energy structure, along with growing energy demand and high pressure for CO_2 emission reduction, poses great challenges for China's future energy structure development. Thus, a hybrid energy system was proposed as a possible solution to utilize China's high carbon energy resources in a green and low carbon way. High carbon energy resources such as coal could be coupled with low carbon energy resources like nuclear energy and renewable energies by the effective integration between materials, energy and information.

In this case, both energy and carbon efficiencies could be highly improved with a reduction of CO_2 emission, and at the same time CO_2 could be used as a feedstock. As a result, hybrid energy system is expected to be a strategic solution to China's future energy structure development. This study addresses this topic by investigating the technical and economic feasibility of different hybrid energy systems.

Based on the above consideration, nuclear-assisted coal-based energy system and wind-assisted coal-based energy system were constructed to utilize the nuclear energy in China's coast area and wind energy in Inner Mongonia, respectively. The results obtained in this study demonstrated the economic and environmental superiority of different hybrid energy systems.

Furthermore, the results clearly indicated that hybrid energy system is, in general, a feasible and preferable alternative solution to China's future energy supply. The research methodology and procedure developed in this project can be used to investigate and identify the most viable hybrid energy system based on China's special energy resource endowment and economic circumstances.

(87d) Efficient Hydrogen-Rich Syngas Production Via Microwave-Induced Gasification of Bio-Char Derived From Agricultural Residues

Xiaoquan Wang, Low Carbon Conversion Center, Shanghai Advanced Research Institute,Chinese Academy of Sciences, Shanghai, China, Hu Luo, Low Carbon Conversion Center, Shanghai Advanced Research Institute, Chinese Academy of Sciences, Shanghai, China, Bo Liao Sr., Low Carbon Conversion Center, Shanghai Advanced Research Institute, Shanghai, China and Yuhan Sun, State Key Laboratory of Coal Conversion, Institute of Coal Chemistry, Chinese Academy of Sciences, Taiyuan, China

Rice straw and corn stover are the major agricultural residues produced in large quantities in China, and most of them are consumed as conventional household fuels for stoves. Until recently, China is strongly encouraging the use of biomass for energy purposes, and the pyrolysis of biomass is an attractive option, for it provides a highly efficient way to produce bio-fuels. The bio-char, as a by-product in the pyrolysis of biomass during bio-oil production, is qualified to be used for syngas production due to its high carbon content. Compared with carbon monoxide, hydrogen energy is not only a clean, highly efficient and promising energy, but also can be used as material for hydrogenation to upgrade bio-oil quality, thus comprehensive and systematic investigation on hydrogen-rich syngas production from bio-char is needed.

A potential problem related to char reactivity, in large gasification systems, is that char particles might exit the system prior to their complete conversion. Main reasons for this incomplete conversion include in uneven temperature distribution in the reactor results in the presence of low temperature zones, which subsequently results in low reaction rates to provide poor conversion rates of the char. As microwave energy is transferred directly to the material that is heated, the temperature inside the material is usually higher than the temperature of the surrounding atmosphere, unlike conventional heating. Microwave radiation is known to have the potential to increase the rate of reaction, selectivity and yield of catalytic heterogeneous reactions. The improvement observed under microwave heating is normally attributed to various thermal effects, although the presence of microplasmas or hot spots within the catalyst bed, which are at higher temperature than the average temperature, may be the main reason for the improvement in gas-solid reactions. Microwave heating offers a number of advantages over conventional heating, such as non-contacting heating, rapid and selective heating of materials, higher flexibility, and reducing processing time. So, it could provide a more homogeneous temperature and random microplasmas in different reaction sites, which increases the conversion rates of the char.

The gasification of rice straw and corn stover chars with different oxidizing agents will be investigated in a fixed-bed reactor in a microwave irradiation environment. This exploratory study focuses on the gas composition, concentration of hydrogen, and the carbon conversion efficiency. Meanwhile, it is also expected to determine the influence of char structure and operating condition on gas yields, and find the role of the inherent catalyst in the reaction, thus lay the foundation for engineering application.

(114a) Simulation of Interfacial Rayleigh Convection in Desorption Process by Lattice Boltzmann Method

Bo Fu[1], Xigang Yuan[2], Yang Wang[1], Shaohui Chen[1] and Aiwu Yang[1], (1)Nanjing Research Institute of SINOPEC Yangzi Petrochemical Company Limited, Nanjing, China, (2)State Key Laboratory of Chemical Engineering, Tianjin University, Tianjin, China

Simulation of Interfacial Rayleigh Convection in desorption process by Lattice Boltzmann Method

Bo Fu[1], Xigang Yuan[2], Yang Wang[1], Shaohui Chen[1], Aiwu Yang[1]

1. Nanjing Research Institute of SINOPEC Yangzi Petrochemical Company Limited, Nanjing 210048, China

2. State Key Laboratory of Chemical Engineering, School of Chemical Engineering and Technology, Tianjin University, Tianjin 300072, China

Rayleigh convection is an interfacial phenomenon usually observed in mass transfer processes. In an interfacial mass transfer process, the concentration gradient of the transferred component builds up in the vicinity of the interface, and the adverse density gradient caused by the concentration gradient may generate Rayleigh instability and Rayleigh convection. It is identified that the interfacial mass transfer rate is effectively enhanced by the Rayleigh convection. The research on it finds many important engineering applications such as absorption, desorption, crystallization, extraction, distillation etc. The previous experimental and theoretical studies on Rayleigh convection covered the observation of the Rayleigh convection patterns, measurement of the enhanced mass transfer rate and prediction of onset of Rayleigh convection. As it is difficult to obtain the detailed velocity and concentration fields near the interface through experimental methods, the numerical simulation becomes the important method to investigate on the interfacial Rayleigh convection in mass transfer process. Most of previous simulations for the Rayleigh convection found in the literatures mostly concentrated on absorption process. Compared with the traditional computational fluid dynamics methods, the lattice Boltzmann method (LBM) has become an alternative and effective approach to simulate fluid flow and transport phenomena on the mesoscale.

In the present study, a 2D time-dependent lattice Boltzmann method (LBM) with a double distribution model was established for simulating the liquid-phase Rayleigh convection in the mass transfer process of iso-propylalcohol(IPA) desorption from aqueous solution. The geometry under study is a horizontal layer with a free upper surface regarded as a gas-liquid interface and a rigid bottom. The interfacial concentration was assumed to be constant, and the interfacial disturbance was implemented by the random perturbation model proposed by Fu Bo et al. (2012) with two random parameters $\boldsymbol{P}$ and $\boldsymbol{C}_D$ denoting respectively the possibility and the magnitude of concentration perturbation at interface. With the identified parameters $\boldsymbol{P}$=0.06, $\boldsymbol{C}_D=10^{-12}$ kg°¤m^{-3}, the simulated results show that the critical onset time of the Rayleigh convection increases gradually as the interfacial concentration increases and then keep a stable value of 22 s (see figure 1). As shown in figure 2, the simulations captured the detailed information of the onset and the temporal-spatial evolution of Rayleigh convection, and gave the concentration contours of typical plume convection patterns which were well consistent with Schlieren pictures obtained by Kurenkova et al.(2003). A mass transfer enhancement factor was defined and evaluated by the simulated results, and the interfacial mass transfer was shown to be effectively enhanced by the Rayleigh convection (see figure 3).

Note: the above work mainly was completed in Tianjin University, and the experiment verification would be implemented in Nanjing Research Institute of SINOPEC Yangzi Petrochemical Company Limited.

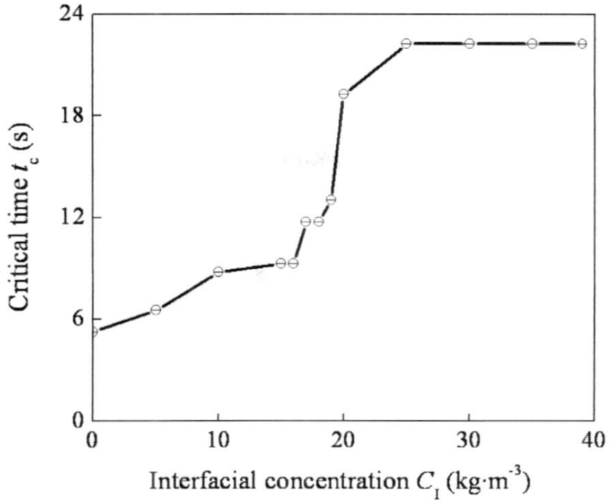

Figure 1 Variation of critical onset times of Rayleigh convection with interfacial concentration in the process of isopropyl alcohol desorption from aqueous solution

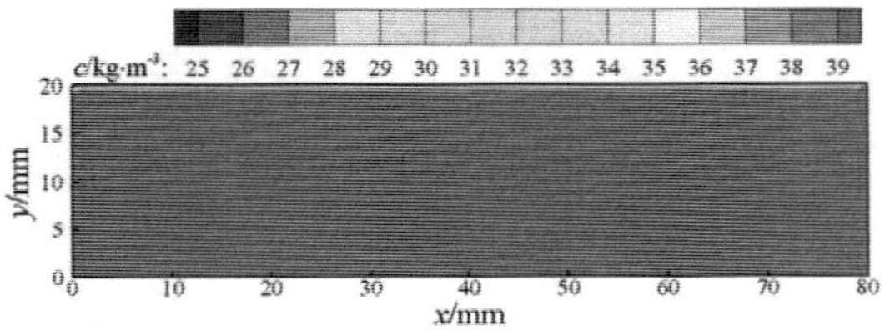

(a) t= 20 s

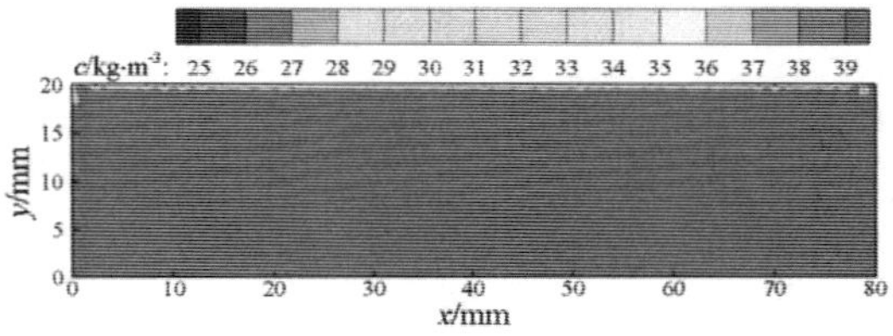

(b) t=25 s

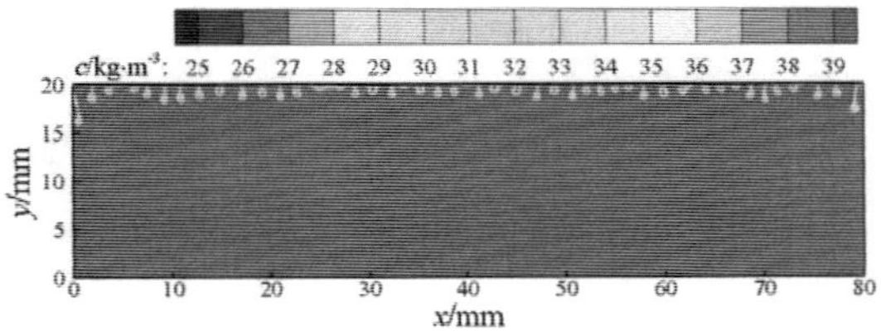

(c) t= 30 s

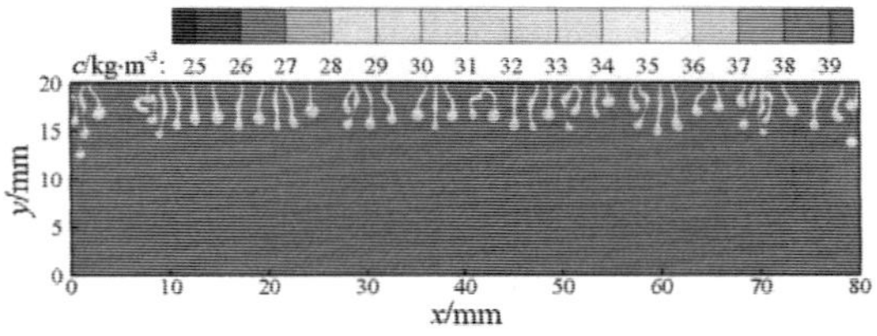

(d) t= 40 s

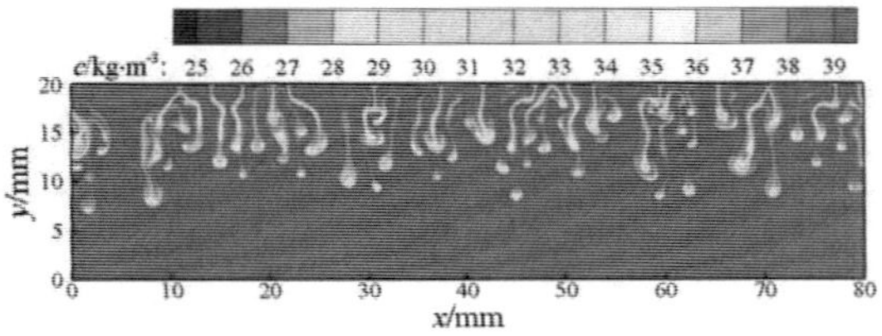

(e) t= 60 s

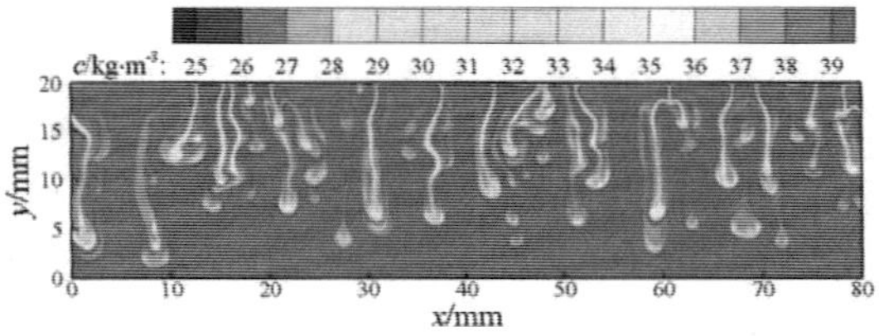

(f) t= 80 s

Figure 2 Transient contours of solute concentration for different time in the process of isopropyl alcohol desorption from aqueous solution

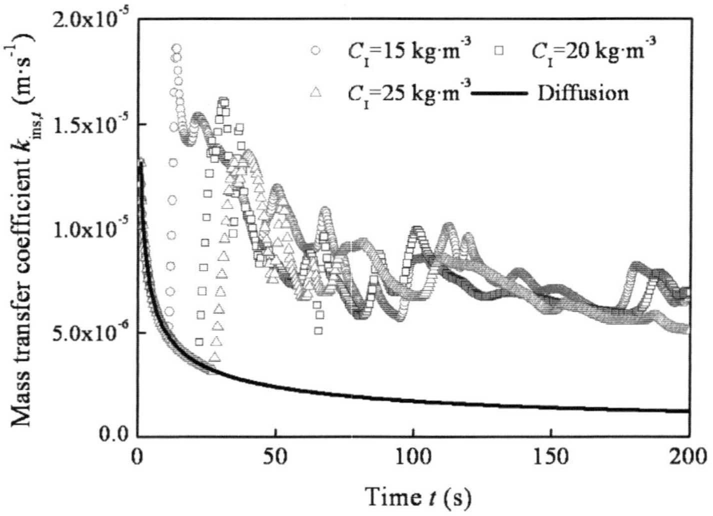

Figure 3 Variation of instantaneous liquid-phase mass transfer coefficient with time for different interfacial concentration in the process of isopropyl alcohol desorption from aqueous solution

Fu Bo, Yuan Xigang, Chen Shuyong, Liu Botan, Yu Kuotsung, 2012. Simulation of Rayleigh Convection in Gas-Liquid Interfacial Mass Transfer Process by Lattice Boltzmann Method. Journal of Tianjin University 45(7), 585-590.

Kurenkova, N., Eckert, K., Zienicke, E., Thess, A., 2003. Desorption-driven convection in aqueous alcohol solution. Lecture Notes Physics 628, 403-416.

(114c) Highly b-Oriented MFI Monolayer Film Synthesized Optical Fiber Chemical Sensor for Analysis of Gas Diffusion

Ruidong Yang, Chemical and Materials Engineering, University of Cincinnati, Cincinnati, OH and Junhang Dong, Chemical Engineering, University of Cincinnati, Cincinnati, OH

A highly b-Oriented monolayer MFI type zeolite film was synthesized on optical fiber end face and a time-resolved microscopic laser refractometry method is established to measure gas diffusivity behavior in zeolite membrane. The monolayer zeolite film was fabricated by direct in situ hydrothermal crystallization upon the cleaved end face of a standard single-model optical fiber. Effect on thin film thickness and surface coverage was investigated by varying crystallization time and composition of the synthesis solution. The films are characterized with scanning electron microscopy and the thicknesses were able to be controlled from 1μm up to 2μm. The zeolite-fiber integrated optical sensor operates by ultra-sensitive and fast-speed monitoring the optical reflectivity change caused by the sorption of gas molecules. Interference spectrums at varied sorbate concentrations\ under equilibrium conditions were analyzed to determine refractive index as function of concentration. The sorbate uptake curves were obtained by measuring single wavelength laser reflectance as a function of time during non-equilibrium dynamic process of adsorption. Gas diffusivity values at different sorbate concentration were derived from each uptake curves mathematically.

LEHCO DESULFURIZER
REMOVE SULFUR AND NITROGEN FROM PETROLEUM COKE
AIChE SPRING 2013
1

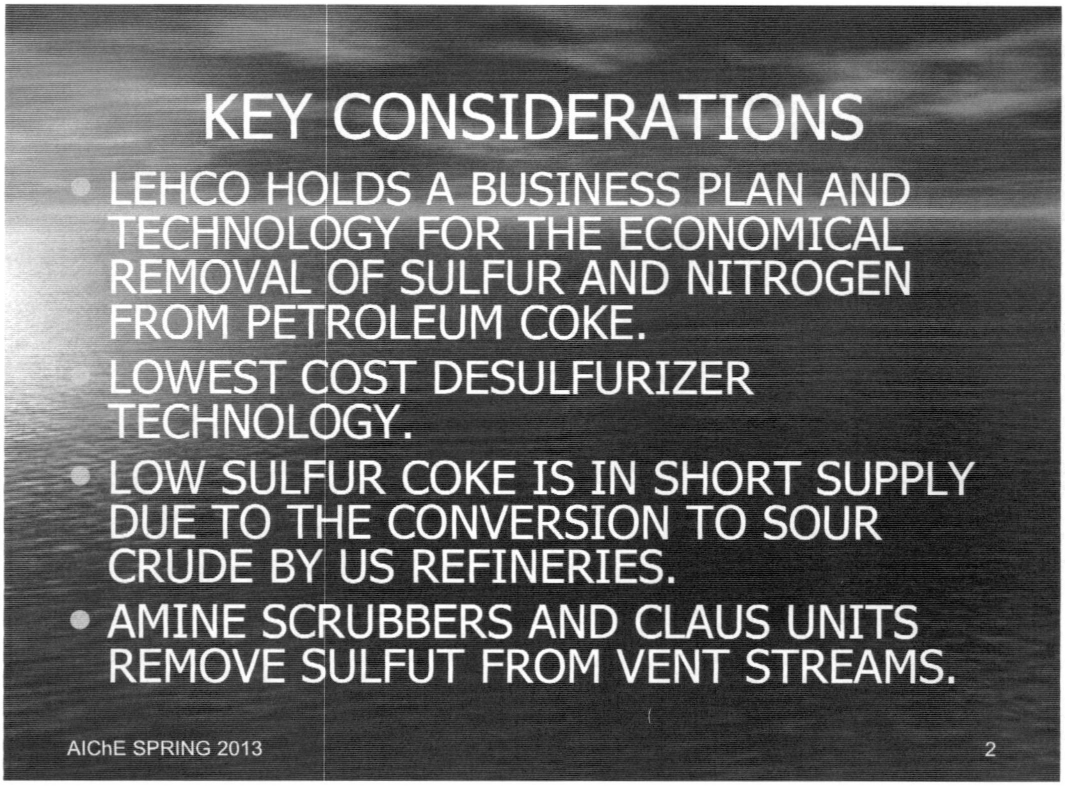
KEY CONSIDERATIONS
LEHCO HOLDS A BUSINESS PLAN AND TECHNOLOGY FOR THE ECONOMICAL REMOVAL OF SULFUR AND NITROGEN FROM PETROLEUM COKE.
LOWEST COST DESULFURIZER TECHNOLOGY.
LOW SULFUR COKE IS IN SHORT SUPPLY DUE TO THE CONVERSION TO SOUR CRUDE BY US REFINERIES.
AMINE SCRUBBERS AND CLAUS UNITS REMOVE SULFUT FROM VENT STREAMS.
AIChE SPRING 2013
2

PRESENT DISPOSITION

- REFINERIES GENERATE LARGE QUANTITIES OF HIGH SULFUR COKE.
- SUCH COKE HAS LIMITED USE IN U.S. MARKETS.
- MOST HIGH SULFUR GREEN COKE IS EXPORTED TO THE FAR EAST.
- SOME U.S. WEST COAST POLLUTION MAY RESULTS FROM SUCH COKE.

LEHCO TECHNOLOGY

- LEHCO HOLDS PATENT 8,372,171 B2.
- SULFUR IS REMOVED BY HEATING THE COKE TO 3000'F IN A COLUMN LINED WITH SELECTED REFRACTORY CAPABLE OF OPERATION AT UP TO 3300'F.
- THE COLUMN OPERATES CONTINUOUSLY UNDER PRESSURE

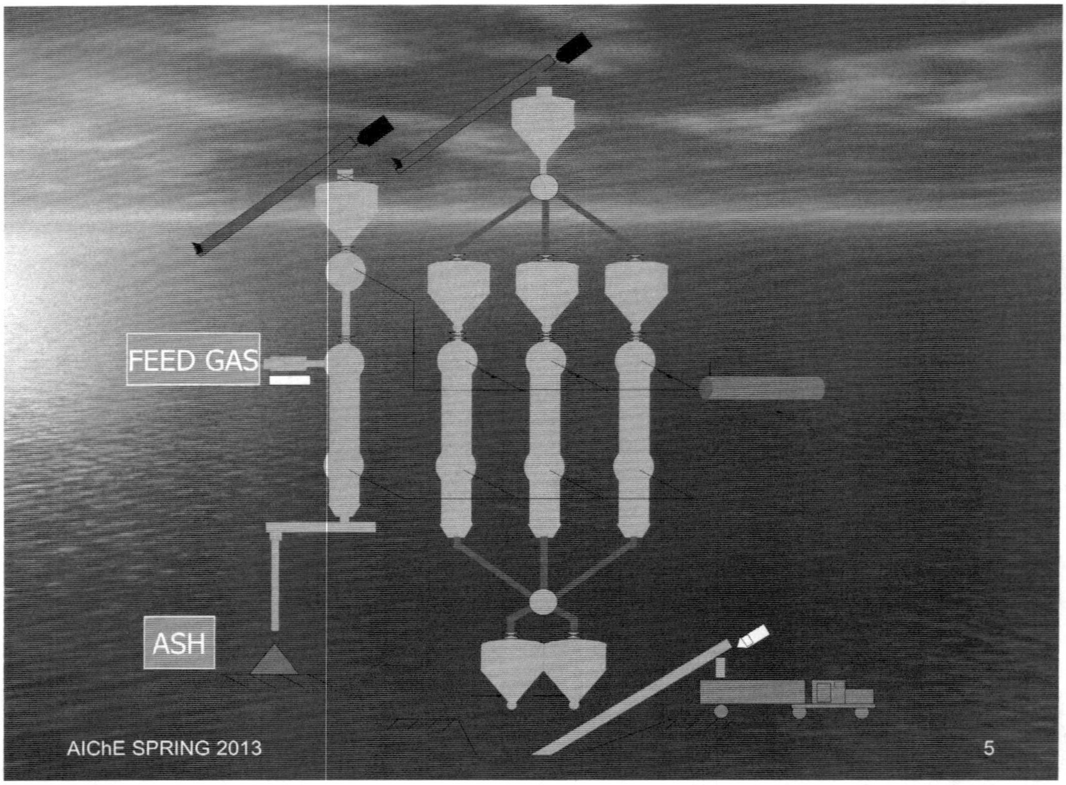
FEED GAS
ASH
AIChE SPRING 2013
5

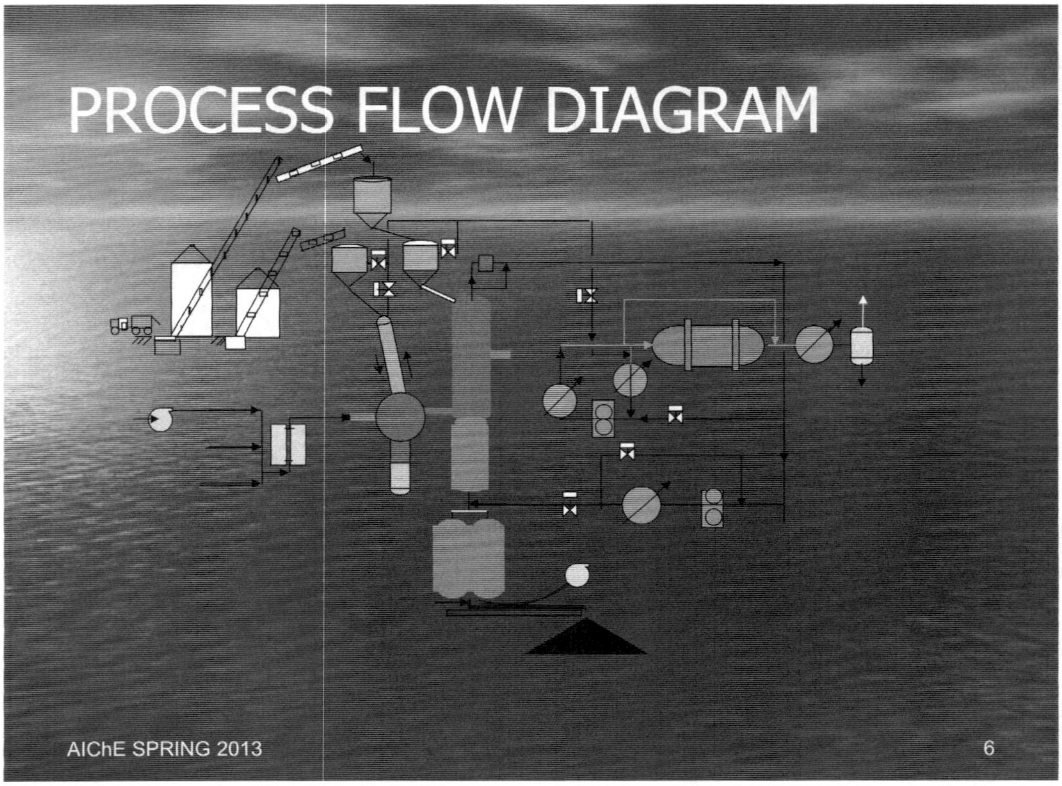
PROCESS FLOW DIAGRAM
AIChE SPRING 2013
6

LEHCO TECHNOLOGY CONT.

- INITIAL PLANS ARE TO DESULFURIZE SHOT COKE.
- FUEL FOR THE PROCESS WILL ALSO BE SHOT COKE.
- DUE TO HIGH PRESSURE THE GAS PASSES UP THROUGH THE COKE BED AS COKE FLOWS DOWN THROUGH THE COLUMN.
- BATCH OPERATION IS ALSO AN OPTION.

AIChE SPRING 2013 7

OTHER CONSIDERATIONS

- CALCINED COKE WILL BE USED IN THE DESULFUIZER.
- GREEN COKE WILL BE USED LATER.
- PRODUCT ALLOCATION WILL BE BASED ON THE DEVELOPING MARKET.
- SOME BLENDING BY USERS IS LIKELY UNTIL THE SUPPLY AND MARKET ARE BALANCED.

AIChE SPRING 2013 8

WHAT HAPPENS TO THE SULFUR

- SULFUR EMERGES AS COS & CS2 GAS.
- AS THE OFF-GAS COOLS IT CAN BE CONVERTED TO H2S BY REACTION WITH H2 GAS.
- COLLECTING THE H2S IS OLD TECHNOLOGY.
- AMINE SCRUBBERS REMOVE H2S.
- H2S GAS IS REGENERATED IN CONCENTRATED FORM BY HEATING.
- REGENERATED H2S GOES TO THE CLAUS.

AIChE SPRING 2013 9

CLAUS UNIT

- CLAUS UNITS CONVERT H2S TO SULFUR
- CLAUS UNITS WERE INVENTED IN LATE 1800'S
- CLAUS UNITS ARE EXPENSIVE.
- DESULFURIZERS SHOULD BE BUILT NEAR AN EXISTING UNIT.
- CLAUS OWNERS WILL HAVE AN EDGE.

AIChE SPRING 2013 10

LEHCO STAFF

- LEHCO HAS TWO PRINCIPLE PARTNERS.
  - LOUIS HERRINGTON, P.E. LOUIS IS THE INVENTOR OF THE PROCESS AND HAS DEVELOPED THE THEMOCHEMISTRY AND REFRACTORYS
  - RICHARD CRISP, HAS MANY YEARS IN CALCINING AND COKE MARKETS, PLUS MANUFACTURE OF FOUNDRY AND FURNACE COKE FROM COAL AND OFF GAS TREATMENT AND RECOVERY.
  - GREAT LAKES SIGNED A NONDISCLOSURE AGREEMENT WITH LEHCO

AIChE SPRING 2013 11

MARKETS

- COKE MARKETS INCLUDE THE FOLLOWING
  - $TiO2$
  - PLASTICS
  - STEEL
  - ANODES
  - CEMENT
- PRESENT MARKETS ARE FORCED TO USE HIGHER SULFUR COKE THAT REQUIRES EXPENSIVE REMEDIATION.

AIChE SPRING 2013 12

REFERENCE

- DESULFURIZATION OF PETROLEUM COKE BEYOND 1600°C by

  CHRISTOPHER A. PAUL; GREAT LAKES CARBON CORPORATION, NOW OXBOW
- LOUIS E. HERRINGTON; LEHCO
  - THE MINERALS, METALS & MATERIALS SOCIETY, 2001
  - AIChE SPRING MEETING 2002

AIChE SPRING 2013 13

SLIDE WILL SHOW GRAPH OF SULFUR VS TEMPERATUE

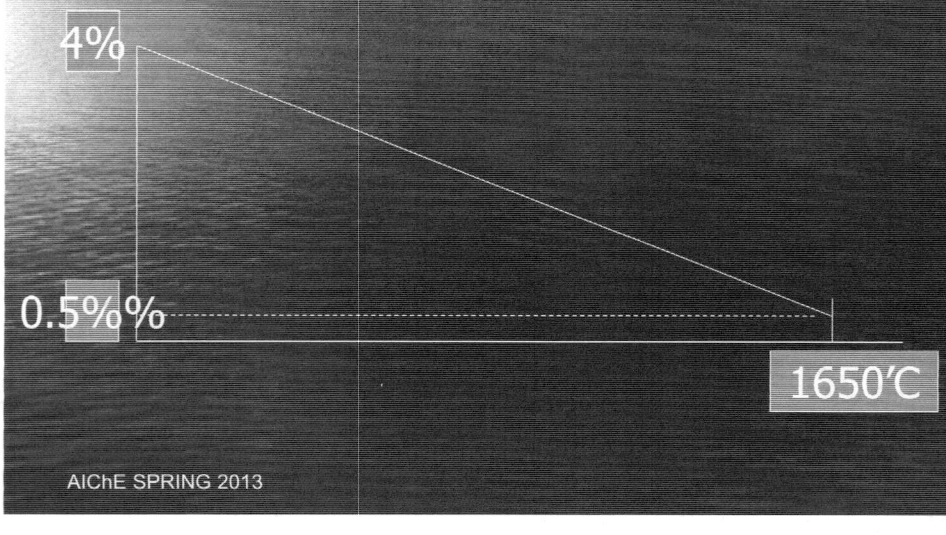

AIChE SPRING 2013 14

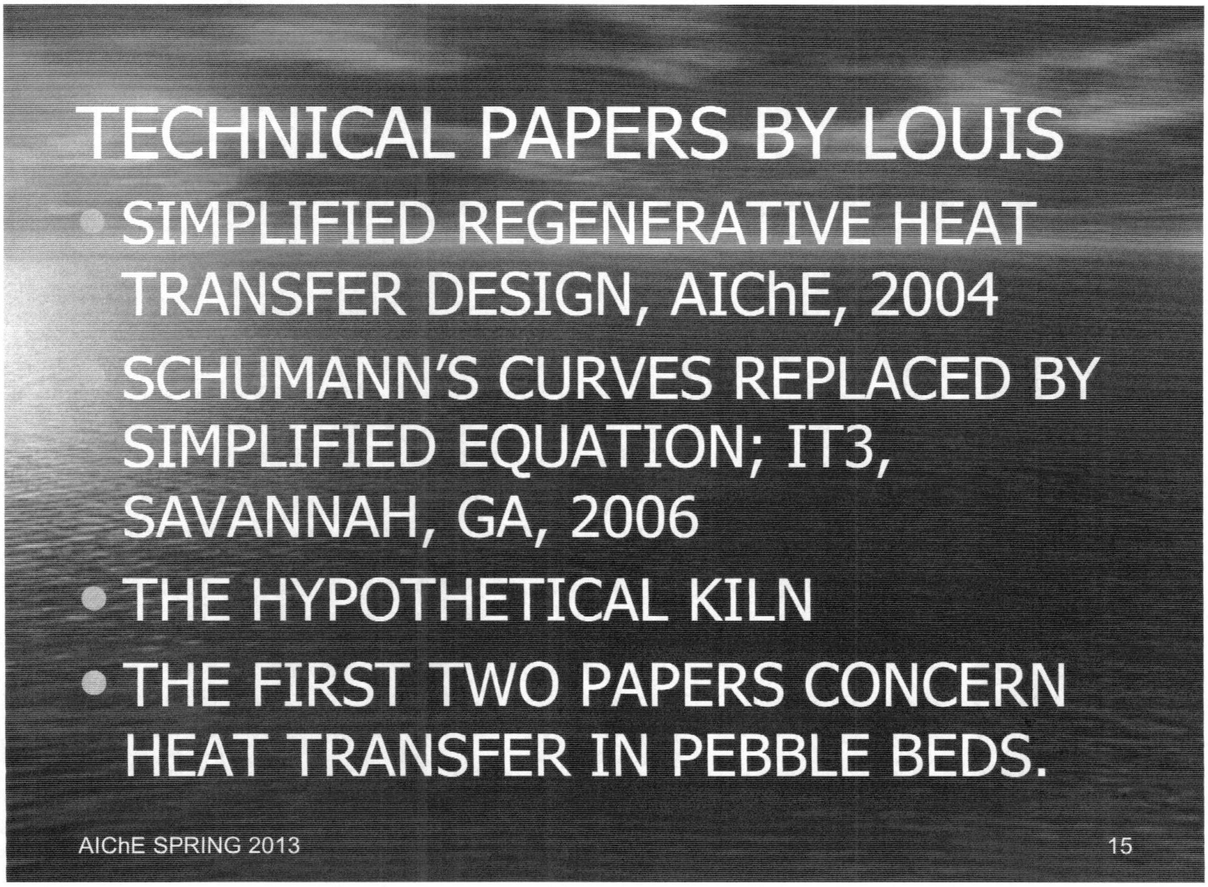
TECHNICAL PAPERS BY LOUIS
SIMPLIFIED REGENERATIVE HEAT TRANSFER DESIGN, AIChE, 2004
SCHUMANN'S CURVES REPLACED BY SIMPLIFIED EQUATION; IT3, SAVANNAH, GA, 2006
THE HYPOTHETICAL KILN
THE FIRST TWO PAPERS CONCERN HEAT TRANSFER IN PEBBLE BEDS.
AIChE SPRING 2013
15

(117a) Refinery Feedstock From Partially Hydrogenated Biomass Pyrolysis Oil

Richard J. French, National Bioenergy Center, National Renewable Energy Laboratory, Golden, CO

Biomass provides a local, relatively secure energy source and may be produced with low or even negative carbon emissions. Therefore biofuels are desirable replacements for petroleum-derived motor-fuels. One method for making biofuels is fast pyrolysis of biomass followed by liquid separation to produce pyrolytic lignin. This might then be hydrotreated to produce a liquid suitable for mixing into a refinery, where it would be further converted into a finished motor-fuel, taking advantage of the existing infrastructure and economies of scale of refineries. The organic compounds in the aqueous by-products could be reformed to produce the hydrogen needed to hydrotreat the lignin.

Pyrolytic lignin contains about half of the energy of the original biomass but consumes less hydrogen than whole pyrolysis oil when hydrogenated because it has much less oxygen. It is much less corrosive than the whole oil because of the removal of carboxylic acids, and is potentially more stable because of the extraction of unstable oxygenates.

To test this hypothesis, several pyrolytic lignins were prepared. Whole pyrolysis oil was mixed with water at 1:1 and 3:1 water-to-oil ratios. The carboxylic acidity in the pyrolytic lignin was reduced to 24 and 10 mgKOH/g-lignin compared to the 81 measured in the whole oil. The carbon content increased from the whole oil to 58-59% on a wet basis in the pyrolytic lignins. These pyrolytic lignins were hydrotreated with hydrotreating catalysts made of sulfided nickel-molybdenum/alumina, platinum/char, or palladium/activated-carbon in a semi-batch 1 L stirred autoclave. The oil was stabilized under hydrogen at 150-280°C, then water and light organics flashed off by partial depressurization. Then hydrodeoxygenation was performed at 340-400°C. The total system pressure was controlled at 70 or 170 bar by addition of hydrogen gas. Gas composition was monitored and condensable vapors were collected throughout experiment. The organic fractions of the condensed vapors and the organic residue remaining in the autoclave were analyzed separately. The residue constituted 65-85% of the organic product.

At hydrodeoxygenation temperatures of 355-400°C under hydrogen at 170 bar total pressure 52–78% of the carbon in the pyrolytic lignin was converted to organic liquid. The organic oxygen in the overall product, consisting of the organic condensates and the organic residue, was low (2-4%). The carboxylic acidity was reduced to 1-7. Volatility by proximate analysis was higher for the organic condensates (97-99%), than the organic residues (93-99%), but was still high overall (93-99%). Most of the products were miscible in heptanes and toluene at 10:1 . However, insufficient miscibility of some liquids, especially the condensate collected during stabilization, reduced the yield of product deemed satisfactory. Nevertheless with all three catalysts over 60% of the carbon in the pyrolytic lignin was converted to liquids with satisfactory properties. At the less severe condition ($T = 340°C$, $p = 70$ bar), the products were not satisfactory.

Analysis of the products by NMR and GC-MS showed that most of the product is aliphatic but the remaining oxygen is associated with the aromatic materials. The products have a high content of cyclic aromatics and aliphatics with one or two rings.

Thus, the reforming of the aqueous liquids to produce hydrogen coupled with the hydrotreating of pyrolytic lignin is a technically feasible process way to convert biomass to hydrocarbon liquids that are promising refinery feedstocks.

(117b) Catalytic Pyrolysis of Model Compounds Using Different Acidic Zeolites

Vaishnavi Srinivasan and Sushil Adhikari, Biosystems Engineering, Auburn University, Auburn, AL

Energy generation from biomass has been a promising alternative to reduce the dependence on energy derived from conventional fossil fuels. The liquid fuel derived from pyrolysis of biomass exhibits certain characteristics that limit its usage as an efficient transportation fuel. Several upgrading methods to produce aromatics and to selectively eliminate oxygenates derived from the pyrolysis process are available, among which catalytic pyrolysis and pre-treatment (torrefaction) process are gaining a lot of research attraction lately.

Fast pyrolysis of biomass using zeolite catalyst has shown to be effective in improving the aromatic production. This study focuses on the aromatic production through catalytic pyrolysis of two model compounds. Biomass constituents—cellulose and lignin were used as model compounds. Pre-treatment (torrefaction) was performed on the model compounds to observe the effect of torrefaction on the pyrolysis products. The influence of Si/Al ratio of zeolite catalyst on selectively eliminating oxygenates during the pyrolysis of raw and torrefied model compounds were investigated. Surface properties of the spent catalyst, coking were analyzed using SEM (Scanning electron microscopy) and Energy dispersion spectrum.

(117c) Synthesis Gas Production by Autothermal Reforming of Crude Glycerol using a Dual Layer Monolith Catalyst

Yujia Liu, Dept of Chemical Engineering and Material Science, Stevens Institute of Technology, Hoboken, NJ and Adeniyi Lawal, Stevens Institute of Technology, Hoboken, NJ

Crude glycerol is the principal by-product of biodiesel production via transesterification of triglycerides. For 3.33 kg of biodiesel produced, approximately 0.3 kg of crude glycerol is formed. As the demand and production of biodiesel increase exponentially, the utilization of the crude glycerol by-product will be of commercial significance. The conversion of crude glycerol to synthesis gas and then to methanol represents an important niche for glycerol processing that could relieve small biodiesel producers from high crude glycerol disposal cost while reducing their dependence on methanol from fossil fuels. The composition of crude glycerol varies depending on the feedstock and the biodiesel production process; glycerol content ranged between 38 and 96%, with containing up to 29% ash and 14% methanol. Those impurities may affect the catalyst activity. New results on the autothermal reforming of crude glycerol using the BASF dual layer monolith catalyst are presented and compared with results of pure glycerol. Heat management requires balancing the exothermic partial oxidation and endothermic reforming reactions in the catalytic autothermal reforming system. The composition of synthesis gas obtained was determined as a function of temperature, steam to carbon ratio, oxygen to carbon ratio and gas hourly space velocity. The optimum operating conditions to produce desired conversion of crude glycerol, yields of H2, and H2 /CO ratio with minimal coke formation were also determined. The Aspen simulation software package was used to determine the equilibrium product composition from major compounds for various reaction conditions. A comparison between process simulation results and experimental data was also made.

(117d) Thermophysical Properties Measurements for Bio-Oil Upgrading Process

Erlin Sapei[1], Juho Autio[1], Piia Haimi[1], Thomas Kohl[2], Timo Laukkanen[2], Ville Alopaeus[1] and Mika Jarvinen[2], (1)Biotechnology and Chemical Technology, Aalto University, Espoo, Finland, (2)Energy Technology, Aalto University, Espoo, Finland

The use of biofuel as a road transportation fuel is increasing throughout Europe and other industrialized countries to reduce greenhouse gas emissions and to improve the security of fuel supply. Bio-oil produced from biomass pyrolysis is renewable liquid fuel and can also be used for production of chemicals. Upgrading of the bio-oil to transportation fuel is technically feasible and needs further development. Thermophysical properties of bio-oil components have recently received attention because these data are required to obtain accurate thermodynamic models of the compounds formed during pyrolysis. However, the experimental data on the thermophysical properties of the bio-oil is very limited in the literature. The objective of our research is to measure the thermophysical properties of the main components existing in the bio-oil using different techniques and newly built apparatuses. The experimental data were used for the estimation of the bio-oil composition in biomass fast pyrolysis. The estimation was done using a mathematical optimization model based on mass and energy balance in GAMS (General Algebraic Modeling System).

(117e) Fractionation and Characterization of Bio-Oils Produced From Microwave-Assisted Pyrolysis of Rice Straw

Hu Luo, Low Carbon Conversion Center, Shanghai Advanced Research Institute, Chinese Academy of Sciences, Shanghai, China, Bo Liao Sr., Low Carbon Conversion Center, Shanghai Advanced Research Institute, Shanghai, China, Yuhan Sun, State Key Laboratory of Coal Conversion, Institute of Coal Chemistry, Chinese Academy of Sciences, Taiyuan, China and **Xiaoquan Wang**, Low Carbon Conversion Center, Shanghai Advanced Research Institute,Chinese Academy of Sciences, Shanghai, China

Bio-oil from biomass pyrolysis is a complex mixture of polar and non-polar compounds, and cannot be used directly as a liquid engine fuel due to its low heating value, high viscosity, chemical instability, and incomplete volatility. The aims of this work are to develop an analytical approach and separation process to determine the chemical composition of bio-oils produced frommicrowave-assistedpyrolysis of rice straw. The bio-oils are supposed to be firstly separated to different fractions which still mixtures using solvent extraction, and then each fraction is analyzed using GC-MS, Gel Permeation Chromatography (GPC) and thermogravimetric (TG) techniques. The bio-oil composition can be described as a mixture of water, volatile acids, aldehydes and ketones, pyrans and furans, sugars, phenyl derivatives, lignin extractives and solids. This method can be used to monitor the main differences in composition of various biomass derived pyrolysis liquids.

Ceramic and Coal: ITM Oxygen for Power Generation with Reduced CO_2-Emissions, Detailed Engineering Study Results

John M. Repasky[1*], *VanEric E. Stein*[1*], *Phillip A. Armstrong*[1*],
Merrill S. Quintrell[2], *Andrew Maxson*[3], L. Michael Bartone[4]

[1.] *Air Products and Chemicals, Inc., 7201 Hamilton Blvd, Allentown, PA, 18195 USA*
[2.] *Electric Power Research Institute (EPRI)*, 1300 West W.T. Harris Blvd, Charlotte, NC 28262
[3.] *Electric Power Research Institute (EPRI)*, 3420 Hillview Avenue, Palo Alto, CA 94304
[4.] WorleyParsons Group, Inc., 2675 Morgantown Rd, Reading, PA 19607
[*] *Correspondence should be addressed to:*
J.M. Repasky, email: repaskjm@airproducts.com, tel: +1 (610) 481-5653
V.E. Stein, email: steinve@airproducts.com, tel: +1 (610) 481-4639
P.A. Armstrong, email: armstrpa@airproducts.com, tel: +1 (610) 481-8754

Prepared for Presentation at the AIChE Spring National Meeting
Topical 5: Emerging Technologies in Clean Energy for the Twenty-First Century
Session on Carbon Capture, Storage and Utilization Routes to Produce Chemicals
San Antonio, TX April 28–May 2, 2013

Abstract

In partnership with the U.S. Department of Energy, an Air Products-led team is developing a new air separation technology - Ion Transport Membrane Oxygen - based on ceramic membranes that selectively transport oxygen ions when operated at high temperature. Under the influence of an oxygen partial-pressure driving force, the electrochemical ITM Oxygen process achieves a high-purity, high-flux separation of oxygen from air. This high temperature process lends itself to integration with advanced power generation processes that require oxygen feedstock, such as Oxy-Coal Combustion and IGCC, as well as traditional industrial applications for oxygen and distributed power.

The team has successfully demonstrated expected performance and operability of commercial-scale modules in a prototype facility that produces up to 5 tons-per-day (TPD) of oxygen. Next phase demonstration unit is in construction, and operation is planned in 2013. This Intermediate Scale Test Unit (ISTU) is designed to produce up to 100 TPD of oxygen integrated with turbo-machinery and co-production of power. Data from the 100 TPD unit will provide the design basis for a much larger plant that could produce 2000 TPD. In parallel, work has also begun to expand ceramic fabrication capacity to support these demonstrations toward commercialization. Previous studies have reported significant capital and operating cost reductions afforded by the ITM Oxygen technology toward traditional gasification, IGCC, and other energy-intensive applications.

In this presentation, we will report the results of detailed engineering and economic study of processes toward reduced CO_2 emissions from power generation cycles. An overview and update of the ITM Oxygen development effort will also be presented, including pilot-scale testing results of commercial-scale ceramic modules, next-phase scale-up designs, and ceramic manufacturing expansion work. A commercialization timeline will be discussed.

Introduction

Worldwide energy use is projected to grow significantly over coming decades as developing economies progress. Coal will likely play a significant role in the production of that energy. To compete with other feed stocks and processes, and with the added complication of potential carbon constraints, power generation from coal will need to use the most advanced technologies available. Two of these are gasification and oxy-combustion which enable high efficiency power production and generate a CO_2-rich by-product stream suitable for capture. However, both gasification and oxy-combustion make use of large quantities of oxygen, which today is sourced from cryogenic air separation facilities. Cryogenic air separation is a reliable technology, but one which constitutes, for example, 15% of the cost of an integrated gasification combined-cycle (IGCC) power plant and a significant fraction of the internal power consumption. Moreover, cryogenic technology is mature and is therefore unlikely to undergo future cost reduction coming from step-out technology improvements. A new air separation process is needed to support the role of coal in the future energy mix and industrial manufacturing.

Background

In the late 1980's, Air Products identified a class of perovskite ceramic materials with high flux and separation selectivity to oxygen that can form the basis for cost-efficient air separation membranes. These materials separate oxygen from air at high temperature in an electrochemically driven process. The oxygen in air is ionized on the surface of the ceramic and diffuses through the membrane as oxygen ions, driven by an oxygen activity (partial pressure) gradient to form oxygen molecules on the other side. Impurities such as nitrogen are rejected by the membrane. In addition, these materials conduct electrons as well as ions, such that no external source of electrical power is required – the entire separation is driven by an oxygen partial pressure gradient across the membrane. Detailed descriptions of the materials and electrochemical processes can be found in references 1-3. The air separation system that results from the use of such ceramic oxygen *ion transport membranes* (ITM's) produces a hot, pure oxygen stream and a hot, pressurized, oxygen-depleted stream from which significant amounts of energy can be extracted.

Because of the requirement to feed a high-pressure oxygen-containing gas to the membrane separator and the production of the hot high-pressure off-gas stream, ITM Oxygen systems advantageously integrate with turbo-machinery-based power cycles. Compressed air, obtained directly from an air compressor or extracted from the compression side of a gas turbine, is heated to 800-900° C, and supplied to the membrane. A portion of the oxygen is extracted across the membrane, and the resulting oxygen-depleted non-permeate off-gas (which can contain sufficient oxygen to support post-combustion, if desired) can be expanded in an expander or combustion turbine to recover useful work. Additional processing in a heat-recovery-steam-generator (HRSG) can result in an overall product mix of oxygen, power, and steam.

Thermodynamic and process economic analyses indicate that the ITM Oxygen process produces oxygen at significantly lower cost than conventional cryogenic processes. The high selectivity to oxygen separation afforded by the membrane and its rapid oxygen flux result in compact membrane devices with little pressure drop on the high-pressure air side. One implementation of an ITM Oxygen membrane in a power cycle is shown in Figure 1. A pressurized air stream is extracted from a gas turbine and fed to the ITM unit after passing through a direct combustor to raise the feed temperature to 800-900° C; oxygen-depleted off-gas from the membrane unit, at substantially the same pressure as the feed air, is reinserted into the combustion turbine for conversion of its thermal and hydraulic energy to shaft work. The compressed air not extracted from the gas turbine goes directly to the combustion turbine (not shown). More details on the integration of ITM Oxygen with gas turbines can be found in reference 4. By contrast, when air is fed to a cryogenic air separation unit (ASU) instead of the ITM Oxygen unit, heat in the compressed air stream is rejected and much of the pressure energy is expended during the oxygen separation.

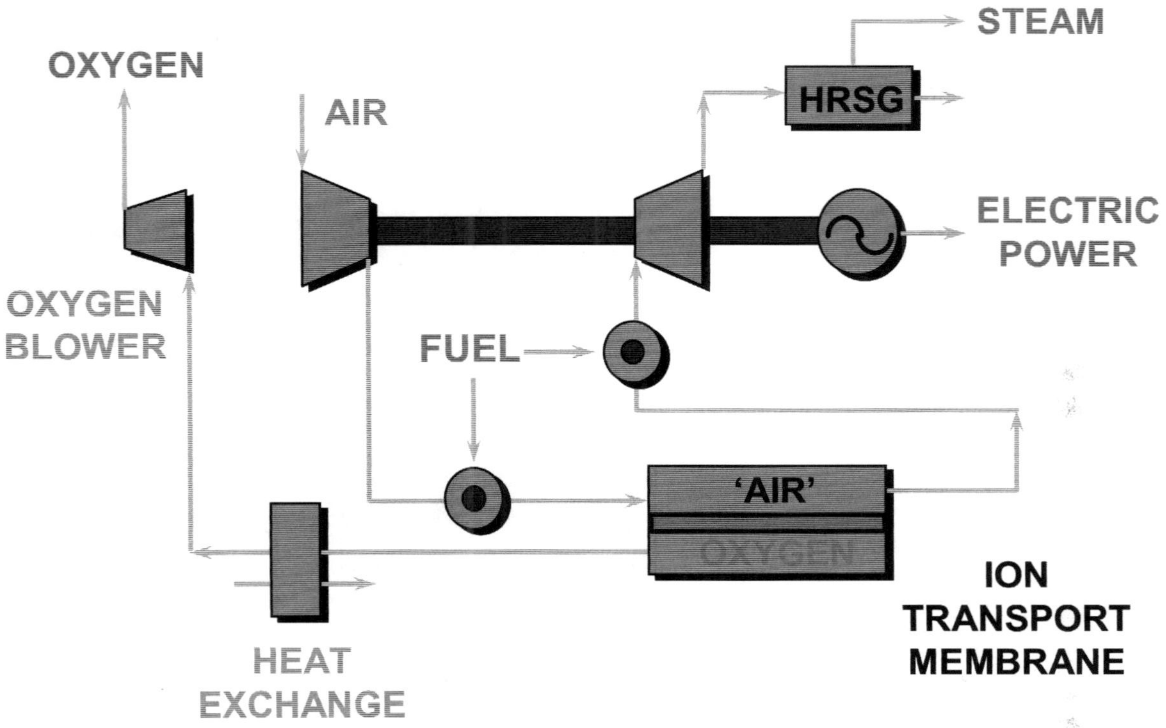

Figure 1. Simplified process flow diagram of an ITM Oxygen unit integrated into a power cycle.

Collaborative Development with the Department of Energy

The potential for co-producing oxygen and power or producing oxygen with relatively little power input has interested the U.S. Department of Energy (U.S. DOE) and others in developing this technology further for advanced applications requiring low cost oxygen. Air Products and the U.S. DOE entered into a Cooperative Agreement in 1998 to develop ITM Oxygen membrane technology in a three-phase program. The goal of the program is to reduce the cost of oxygen significantly relative to conventional cryogenic distillation-based technology.

The development effort has focused on identifying key technical objectives and addressing them comprehensively with a multi-disciplinary, multi-organizational team approach. The program addresses the following areas, among others: materials development, ceramic processing, integration of membrane systems with turbo-machinery, mechanical design, ceramic and system reliability, oxygen production performance, and process safety.

Phase 1 of the program, which focused on the technical feasibility of the approach, was completed in 2001. A material was chosen from a class of perovskite ceramics as the basis for further scale-up. The material has a combination of properties sufficient to meet commercial requirements for performance and operating life, including high oxygen flux, good material strength at high temperature, and resistance to system contaminants such as sulfur. In addition, the material is amenable to standard ceramic processing techniques which facilitates the design, scale-up, and manufacture of multi-layer, planar wafer structures.

A schematic of the planar wafer design is shown in Figure 2. Four wafers are shown joined to a common oxygen withdrawal tube. Each wafer consists of two thin outer membrane layers through which the oxygen ions diffuse. The thin layers on the top and bottom of each wafer are supported by a porous layer which is itself supported by a slotted layer. Hot, high-pressure air flows between the wafers. Oxygen passes from the air outside each wafer, through the thin outer layer, through the pores of the porous layer and into the slots of the innermost layer. Oxygen is collected at the center of each wafer in a tubular region formed by the joined wafers, and passes out of the module through a ceramic tube sealed to a metal pipe. The high-pressure air on both sides of each wafer creates compressive stresses within the ceramic which stabilize the wafer. The planar design makes for a very compact separation device, while facilitating good gas phase mass transfer. All of the layers are made of the same ceramic material, and therefore expand and contract together during temperature changes.

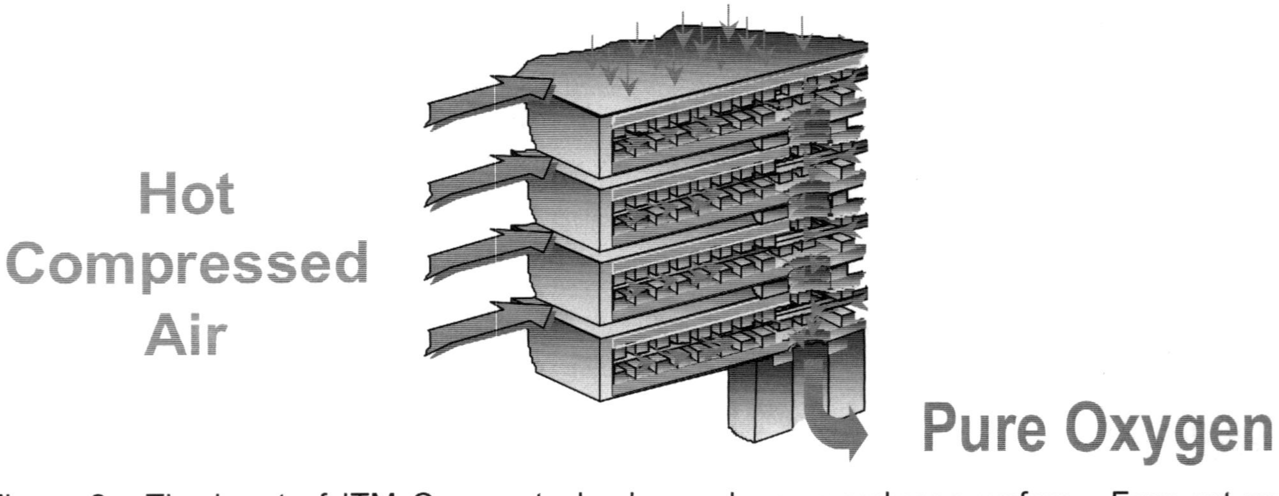

Figure 2. The heart of ITM Oxygen technology: planar membrane wafers. Four cut-away wafers are shown on a common oxygen withdrawal tube. Hot compressed air flows between each adjacent pair of wafers; oxygen is withdrawn at the bottom of the schematic.

The planar wafers were scaled to their full commercial dimensions and produced in volume on a pilot production line using standard ceramic tape-casting technology. The production activities established the feasibility of achieving low-cost production required to meet overall economic targets. Multi-wafer modules were constructed, and the modules were operated under commercially relevant high-temperature, high-pressure conditions in pilot-scale experiments. Oxygen production exceeded commercial flux and purity targets. Detailed process economic studies re-confirmed the overall, market-disruptive benefits of the technology (see reference 5).

The commercial concept of an ITM Oxygen separation vessel is shown in Figure 3. An array of multi-wafer modules (in green) is contained in a common flow duct and connected through a series of manifolds to an oxygen header below. Each commercial-scale module produces about one (1) ton-per-day (TPD) of oxygen, and many modules are arrayed in parallel and series to meet the production requirements of a large tonnage oxygen plant.

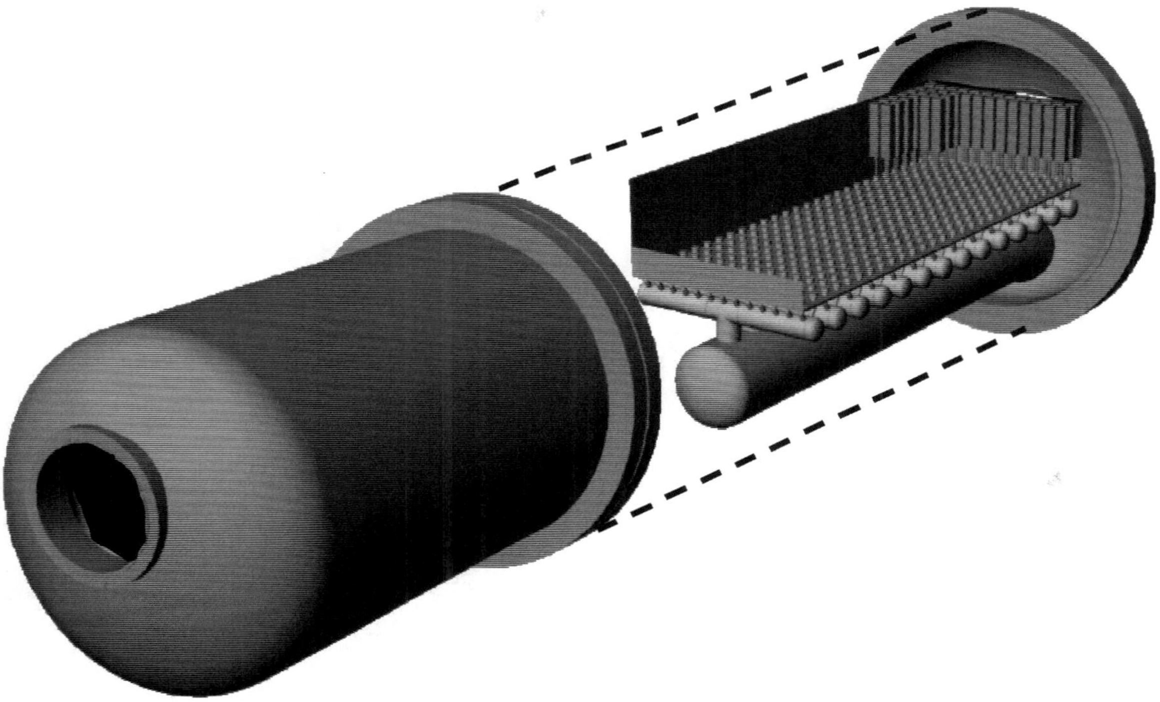

Figure 3. Schematic of commercial ITM Oxygen vessel showing ITM Oxygen modules (in green) arranged in a flow duct and connected to a common oxygen header. The vessel is shown only partially full of modules to expose the oxygen gas manifold tubes below.

In Phase 2, commercial-scale modules, capable of producing up to 0.5 TPD of oxygen were built by co-joining multiple wafers to form a unified ceramic device (see references 6, 7, 8, 9, and 10). In Phase 3, the ceramic architecture was further improved and commercial-scale modules were further scaled up to produce up to 1 TPD of oxygen. The modules are fitted with a terminating end cap and ceramic oxygen product pipe. All ceramic joints are composed of the same ceramic material as the wafers, thus minimizing the potential for differential stresses caused by non-uniform expansion across the body of the device. Figure 4 shows four 1 TPD ITM Oxygen modules awaiting testing after fabrication.

The pilot plant or Subscale Engineering Prototype (SEP) was designed, constructed and commissioned at Sparrows Point, Maryland during the latter portion of Phase 2 and operation has extended through Phase 3. This unit features a prototype ITM pressure vessel which holds six commercial scale modules in a 2-by-3 array. Nominal membrane operating conditions for the unit are 800-900° C, 200-250 psig feed pressure, and <1 atm permeate pressure. The SEP is equipped with sufficient flow capability to simulate feed gas velocities anticipated in commercial service. An aerial photograph of the SEP is shown in Figure 5.

Figure 4. Commercial-scale ceramic ITM Oxygen modules. Each module is capable of producing 1 TPD oxygen.

Figure 5. Aerial photograph of the ITM Oxygen SEP at Sparrows Point, Maryland.

Air Products carried out the first experiment in the SEP in January 2006, and the SEP has operated for a total on-stream time of over 1000 days. Multiple runs have been performed in order to investigate a number of operating parameters and support further scale-up and commercialization of ITM Oxygen technology. Expected values of purity and flux have been measured during multiple runs, including oxygen purities of 99.9% (in excess of the commercial purity target of 99.5% oxygen and well in excess of the purity of 95% required for gasification and oxy-combustion power applications). Various types of ceramic-to-metal seals have been tested, as have a range of startup and shutdown schedules designed to maintain ceramic reliability while reducing startup and shutdown times. Multiple banks of modules were tested, and their ability to withstand rapid transients in process conditions was studied. Full scale 1 TPD modules were tested, as were various mechanical components designed for use in the next scale-up unit. Contaminant getter systems and auto-shutoff isolation valves have been developed and tested toward reliable extended operation of commercial ITM Oxygen systems. These results have been published previously, and references 6-10 contain more detailed discussion of the SEP configuration and test results.

Process Development

Numerous process and economic studies have been conducted of ITM Oxygen; ITM Oxygen has consistently shown excellent economic performance across multiple applications where oxygen and power are needed. ITM Oxygen technology integrates well with power generation cycles, and studies have shown that ITM Oxygen, optimally configured, has the potential to result in 25% or greater reduction in the cost of oxygen compared to conventional cryogenic oxygen plants in base load clean power generation applications. ITM Oxygen shows attractive savings toward high pressure IGCC and decarburized fuel applications relative to a cryogenic oxygen plant (pumped LOX cycle), and shows even greater power savings toward low pressure oxygen enrichment and oxyfuel applications relative to a cryogenic oxygen plant (LP cycle); see references 6-10. An extensive study on the use of ITM Oxygen for oxy-firing process boilers and heaters in a refinery to avoid CO_2 emissions was published previously in reference 11. ITM Oxygen is also attractive in other industries, particularly those using high temperature processes or combustion (e.g. steel, non-ferrous metals, hydrometallurgy, pulp and paper, chemicals, petrochemicals, glass, and cement). Industries with moderate-scale tonnage oxygen requirements may provide opportunity for early application of ITM Oxygen technology.

ITM Oxygen integrates advantageously with well-known approaches for CO_2 capture, including oxyfuel combustion capture and pre-combustion capture routes, and provides benefits in power generation with reduced CO_2 emissions. Previous work has shown ITM Oxygen to integrate well with Oxy-combustion (see reference 12) and IGCC (see references 13-15). General oxyfuel combustion, already in use in various industries such as glass manufacture, may be employed in a number of applications to reduce CO_2 emissions. Pre-combustion capture is optimally suited for applications where carbon-free fuel (e.g. hydrogen) is produced from hydrocarbon feedstock, for example, in refining or gasification applications such as gas-to-liquids (GTL), coal-to-liquids (CTL), coal gasification to poly-generation chemicals, and IGCC power generation. Different ITM Oxygen process configurations appear optimal toward different applications, and extensive study of ITM Oxygen process cycle development toward power generation with reduced CO_2-emissions was published previously in reference 16; additional ITM process optimization fundamentals are discussed in references 17-19.

Further detailed engineering studies are being conducted with input from power industry companies toward large-scale power generation with carbon capture, including Oxy-coal Combustion with input from boiler manufacturers and Integrated Gasification Combined Cycle (IGCC) with input from turbine manufacturers. Below we present details from two recently completed studies contrasting the performance of an ITM Oxygen-based oxygen supply with that of a cryogenic Air Separation Unit (ASU)-based oxygen supply in coal-based power generation applications with carbon capture. These studies focus on oxygen supply and augment the growing body of information that multiple authors have contributed in the area of power generation with carbon capture.

i. <u>Oxy-coal Combustion Power Generation with Carbon Capture</u>

A detailed engineering study of a green-field 800 MW (gross) oxy-combustion supercritical pulverized coal steam boiler with carbon capture, purification, and compression compared the supply of oxygen from ITM Oxygen with traditional cryogenic air separation units (ASU's). The process was designed and its performance assessed using an Aspen process simulator and additional proprietary software. Steam turbine throttle conditions were 3515 psia (242 bara) / 1110°F (600°C) / 1150°F (621°C). Typical of large-scale boiler design and oxy-combustion technology, the distribution of primary air, secondary air, and flue gas recycle (including pre-heat and reheat economizers) was used to manage flame temperatures and heat release profiles in the boiler (see references 20-22). Flue gas processing included a bag house for fly ash removal and lime slurry based flue gas desulfurization. After initial processing, the wet flue gas was divided for recycle to the boiler (nominally 70%) and sequestration (nominally 30%). Processing of the raw flue gas for sequestration included initial compression to 510 psia (35 bara), drying and CO_2 purification, and final compression of the CO_2 product to 2215 psia (152.7 bara) for sequestration. The scope of the study concluded at the battery limit of the Oxy-coal Combustion facility; CO_2 transport, storage, and monitoring were not considered. In order to focus the study on comparison of oxygen supply technologies, the power plant models used a common design basis except when limits posed by the individual technologies warranted the use of different design bases (e.g. oxygen purity). A block diagram of the Oxy-coal Combustion power generation facility with carbon capture, purification, and compression is shown in Figure 6.

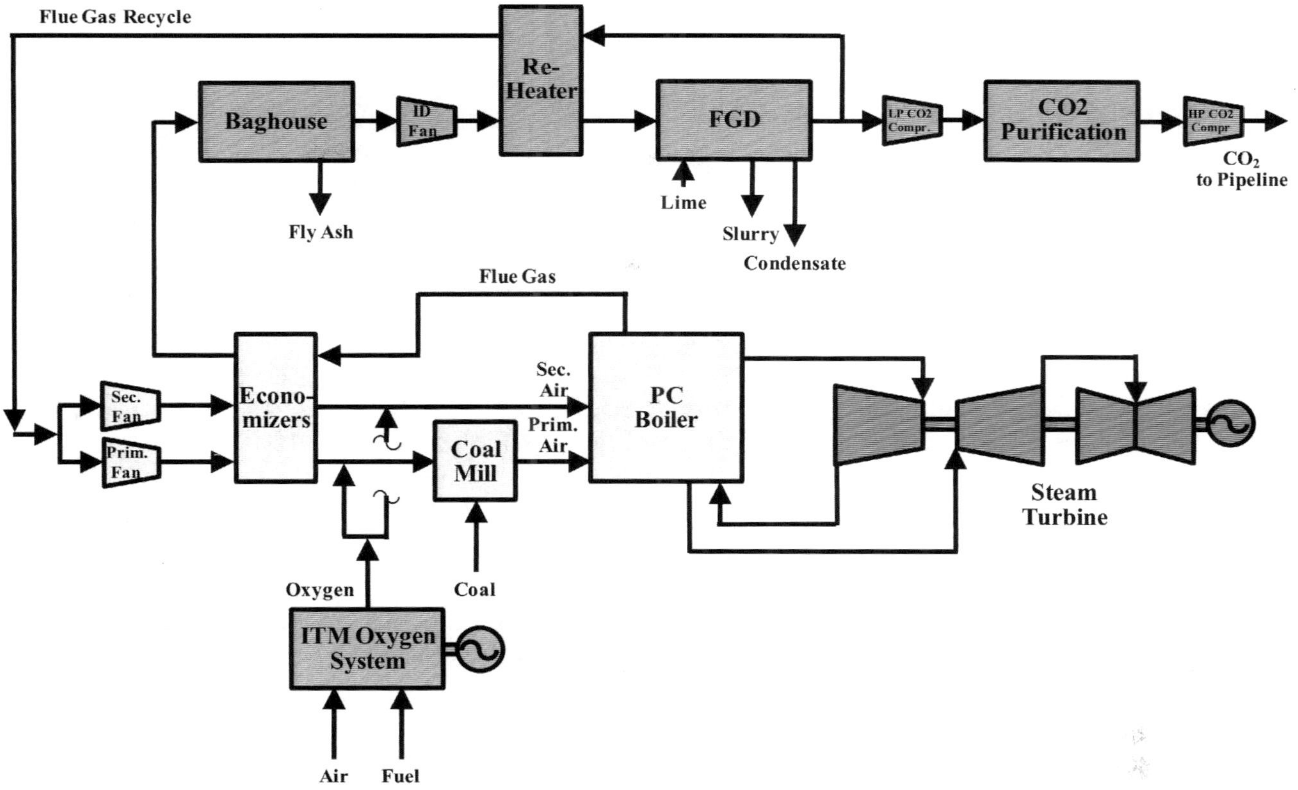

Figure 6. Block flow diagram of the Oxy-coal Combustion power generation facility with carbon capture, showing oxygen supply by ITM Oxygen.

A matrix of six Oxy-coal Combustion cases was fully designed, performance-modeled, cost estimated, and evaluated, including three cases using 95 mol% oxygen provided by cryogenic ASU's and three cases using 99.5 mol% oxygen provided by ITM Oxygen systems.

- An initial cryogenic ASU case targeted 90% CO_2 capture without CO_2 purification. Essentially 100% CO_2 capture was achieved. However, the resulting oxygen concentration in the flue gas (after drying) was about 6 mol%, and it was recognized that further CO_2 purification would be required for CO_2 transport and storage. Thus, a relatively simple two-stage flash CO_2 purification system was introduced into subsequent study cases, and all subsequent cases included both CO_2 drying and two-stage flash purification.
- A second cryogenic ASU case targeted 90% CO_2 capture and included both CO_2 drying and purification, reducing the oxygen concentration in the CO_2 product to less than 2 mol% oxygen. CO_2 purification reduced the CO_2 capture percentage to 93%. The remaining Oxy-coal Combustion cases also include CO_2 drying and purification, and comparative performance of the Oxy-coal Combustion cases is shown relative to this base case in Table 1 (below).
- A third cryogenic ASU case was designed to achieve 73% CO_2 capture, for direct comparison against the direct fired ITM cases at 73% CO_2 capture (below). In this case, a portion of raw flue gas was simply vented and savings recognized in CO_2 purification and compression. In practice, once the investment has been made to oxy-fire the boiler it would appear advantageous to maximize CO_2 capture, and this partial capture case may not be particularly attractive.

- The fourth case introduced ITM Oxygen technology for the supply of oxygen (and eliminated the cryogenic ASU's). A single stage, direct fired ITM Oxygen cycle was evaluated in which a portion of the heat provided to the feed air to the ITM Oxygen unit was provided by combustion of natural gas. In this case, the combustion products were introduced directly into the ITM Oxygen process stream, before and/or after the ITM Oxygen units, in "direct fired" mode, and the ITM Oxygen units were operated at a single oxygen pressure, thus classifying the ITM Oxygen cycle as "single stage, direct-fired." In this case, CO_2 emissions associated with the natural gas combustion were left un-captured, resulting in overall 73% CO_2 capture from the Oxy-coal Combustion power generation facility. ITM Oxygen technology showed significant improvement in net plant efficiency of 7.5 percentage points (%pts) higher than the cryogenic ASU base case at 93% CO_2 capture and 6.5 %pts higher than the comparable cryogenic ASU case at 73% CO_2 capture. A simplified example configuration of a single stage, direct fired ITM Oxygen cycle is shown in Figure 7.
- The fifth study case evaluated a two stage, direct fired ITM Oxygen cycle (see reference 23). A first ITM Oxygen stage was designed for bulk oxygen separation at sufficient back-pressure (2.4 psig) to deliver oxygen product to the boiler without oxygen product cooling and compression, and a second ITM Oxygen stage was designed to operate at vacuum for high oxygen recovery. Oxygen product streams from the two separation stages were blended and provided to the boiler system at mixed temperature 1000 F (538 C). The use of hot oxygen in boiler design, including recycle flue gas optimization, was based on input from boiler manufacturer. The export of valuable heat energy within the oxygen product (instead of being recovered in expensive equipment) from the high temperature ITM Oxygen process was shown to reduce the coal requirement to the boiler and further improve the net plant efficiency of the Oxy-coal Combustion power generation facility by another 1.1 %pts; such integration is fundamentally not possible from typical cryogenic ASU's which produce cold oxygen product. The elimination of the oxygen vacuum blower on the first ITM Oxygen separation stage also has potential to improve reliability. Again, CO_2 emissions associated with natural gas combustion to heat feed air to the ITM Oxygen units were left un-captured, resulting in overall 73% CO_2 capture from the Oxy-coal Combustion power generation facility. A simplified example configuration of a two stage, direct fired ITM Oxygen cycle is shown in Figure 8.

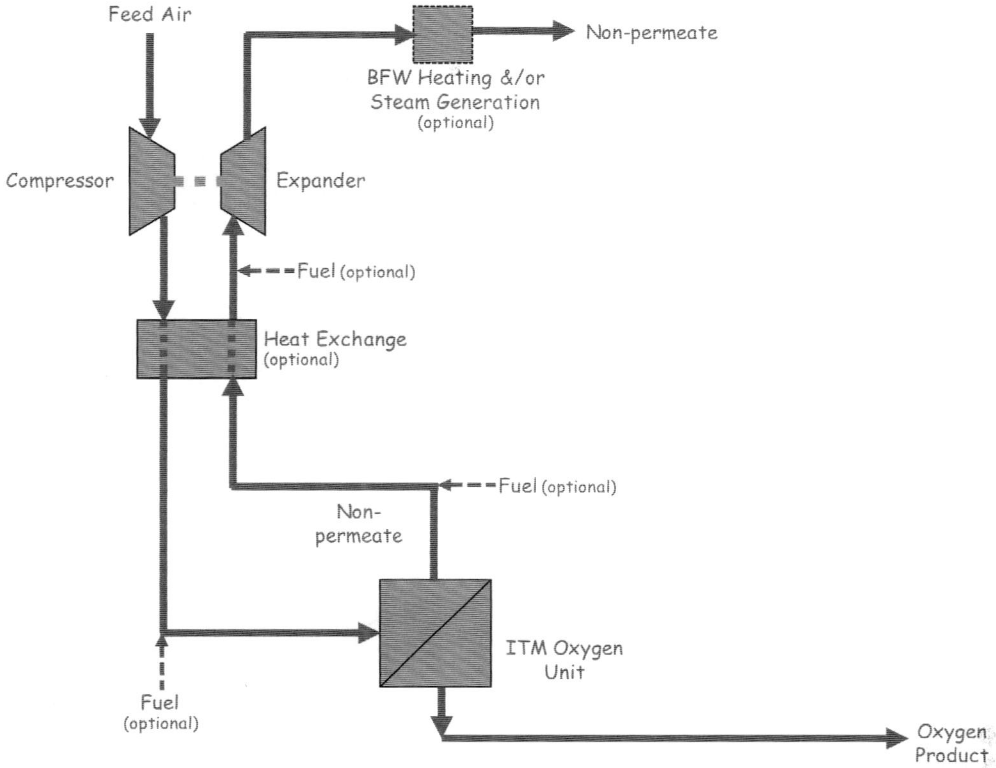

Figure 7. Simplified example configuration of a single stage, direct fired ITM Oxygen cycle.

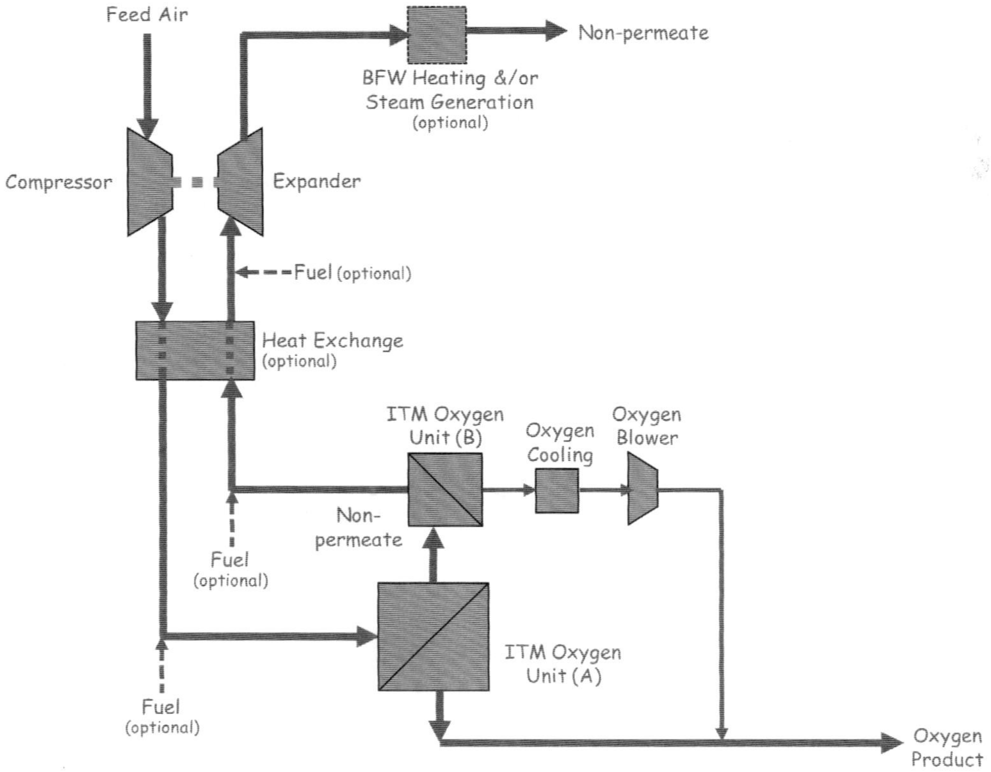

Figure 8. Simplified example configuration of a two stage, direct fired ITM Oxygen cycle.

- The sixth study case targeted 90% CO_2 capture using a two stage, indirect oxy-fired ITM Oxygen cycle. In this case, a portion of the oxygen product from the ITM Oxygen units was consumed within the ITM Oxygen cycle to support indirect oxy-combustion of natural gas to provide heat to the feed air; the resulting flue gases were routed to the power generation facility's CO_2 purification and compression equipment. Overall 90% CO_2 capture was achieved from the Oxy-coal Combustion power generation facility, at the expense of both higher capital investment and net plant efficiency reduction (relative to the direct fired ITM Oxygen cycle at 73% CO_2 capture). Nonetheless, ITM Oxygen technology showed improvement in net plant efficiency of 2.5 %pts higher than the comparable cryogenic ASU base case at 93% CO_2 capture. A simplified example configuration of a two stage, indirect oxy-fired ITM Oxygen cycle is shown in Figure 9.

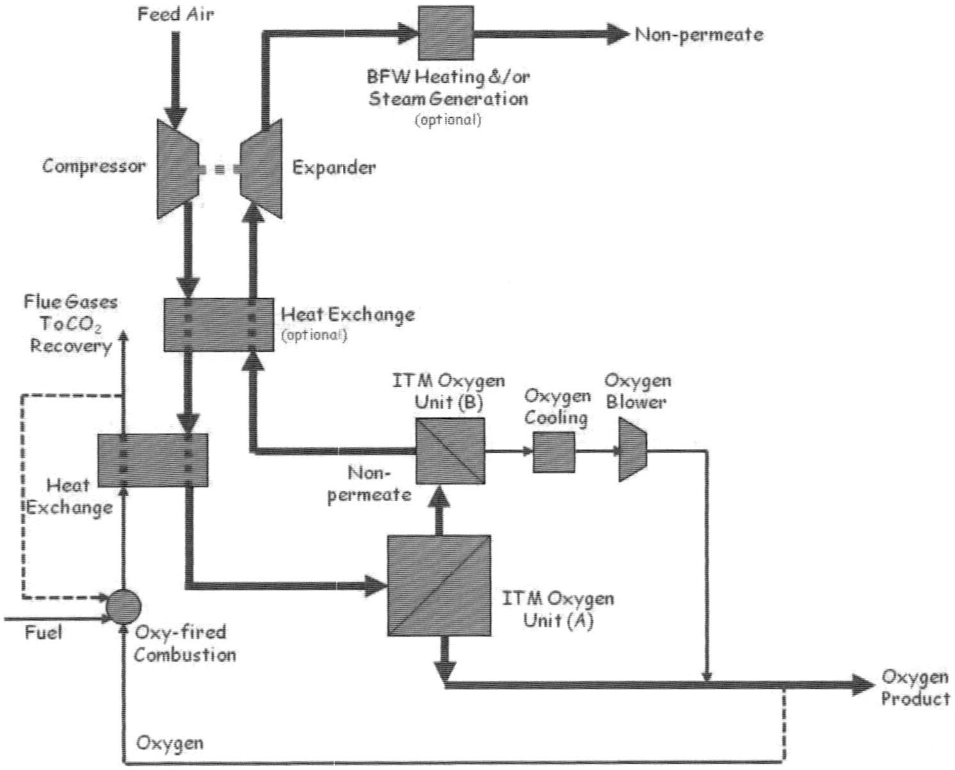

Figure 9. Simplified example configuration of a two stage, indirect oxy-fired ITM Oxygen cycle.

Complete plant designs and cost estimates were generated for each of the six Oxy-coal Combustion Power Plant with Carbon Capture cases described above, and the Levelized Cost of Electricity (LCOE) was calculated over an estimated 30 year plant life.

Summary of the Oxy-coal Combustion Power Generation with Carbon Capture study results is presented in Table 1. At U.S. DOE target 90% CO_2 capture, the ITM Oxygen case (case Oxy-6) showed 2.5 %pts of plant efficiency advantage, 18% lower specific capital cost, and 5% reduction in levelized cost of electricity relative to the cryogenic ASU case (case Oxy-2). The study further showed significant additional advantage of ITM Oxygen technology at reduced 73% CO_2 capture, where the ITM Oxygen case (case Oxy-5) showed 7.6 %pts of efficiency advantage, 34% lower specific capital cost, and 20% reduction in levelized cost of electricity relative to the cryogenic ASU case (case Oxy-3). Regardless of the CO_2 capture percentage, the ITM Oxygen technology cases showed significant increase in net power output, raising the facility power production levels back to the nominal boiler rating without carbon capture.

Table 1. Oxy-coal Combustion Power Generation with Carbon Capture study results.

Oxygen Technology	Cryogenic ASU			ITM Oxygen		
Case	Oxy-1	Oxy-2	Oxy-3	Oxy-4	Oxy-5	Oxy-6
Steam Turbine Power (MW)	800	800	800	800	800	800
ITM Oxygen Net Power (MW)	0	0	0	361	370	177
Total Gross Power (MW)	800	800	800	1161	1170	977
Total Auxiliary Load (MW)	260	267	247	136	132	150
Net Plant Power Output (MW)	540	533	553	1025	1039	827
Net Plant Power Output (MW, Δ%)	+ 1%	base	+ 4%	+ 93%	+ 95%	+ 55%
Net Plant Efficiency, HHV (Δ%pts)	+ 0.3 %pts	base	+ 1.0 %pts	+ 7.5 %pts	+ 8.6 %pts	+ 2.5 %pts
CO2 Captured (kg CO_2/MWH-net)	1,098	1,035	783	548	525	792
CO2 Emitted (kg CO_2/MWH-net)	0	78	289	200	197	89
CO2 Capture Percentage	100%	93%	73%	73%	73%	90%
Total Plant Cost ($, Δ%)	- 1%	base	- 1%	+ 20%	+ 23%	+ 27%
Specific TPC ($/KW-net, Δ%)	- 2%	base	- 5%	- 38%	- 37%	- 18%
Levelized Cost of Electricity ($/MWH, Δ%)	- 2%	base	- 5%	- 24%	- 24%	- 5%

It is noteworthy in the Oxy-coal Combustion Power Generation study results discussed above that typical pulverized coal boiler power generation facilities are based on Rankine cycle steam turbine power generation, and do not include high efficiency gas turbine equipment. While the ITM Oxygen process is fundamentally based on a Brayton cycle, machinery selection for the ITM Oxygen cycles studied was maintained on equivalent basis with equipment selection for the cryogenic ASU cases – using traditional intercooled feed air compression for both cases and separate hot gas expander equipment for the ITM case. Integration of the ITM Oxygen cycle with more efficient gas turbine equipment has potential to further improve cycle efficiency.

ii. <u>Integrated Gasification Combined Cycle (IGCC) Power Generation with Carbon Capture</u>

Detailed engineering study of a nominal 765 MW (gross) coal-based IGCC power generation facility using F-class gas turbine technology with carbon capture compared the supply of oxygen from ITM Oxygen with traditional cryogenic air separation units (ASU's). Raw syngas was generated in a gasification island based on GE radiant quench gasification technology and shifted in two stages of sour shift reactors. Mercury was removed using activated carbon. CO_2 and H_2S were separated from the shifted syngas in a double-stage Selexol unit, with the H_2S being sent to Claus and tail gas treating units and the CO_2 product being compressed to 2215 psia (152.7 bara) for sequestration. The resulting hydrogen-rich fuel was used in the gas turbine to generate electrical power (and to heat feed air in the ITM Oxygen cases). A heat-recovery-steam-generator (HRSG) with three-pressure steam system was included on the gas turbine exhaust, and additional power was generated via steam turbine using steam produced from the HRSG and syngas cooling. The scope of the study concluded at the battery limit of the IGCC facility; CO_2 transport, storage, and monitoring were not considered. In order to focus the study on comparison of oxygen supply technologies, the power plant models used a common design basis except when limits posed by the individual technologies warranted the use of different design bases (e.g. oxygen purity). A block diagram of the IGCC power generation facility with carbon capture and compression is shown in Figure 10.

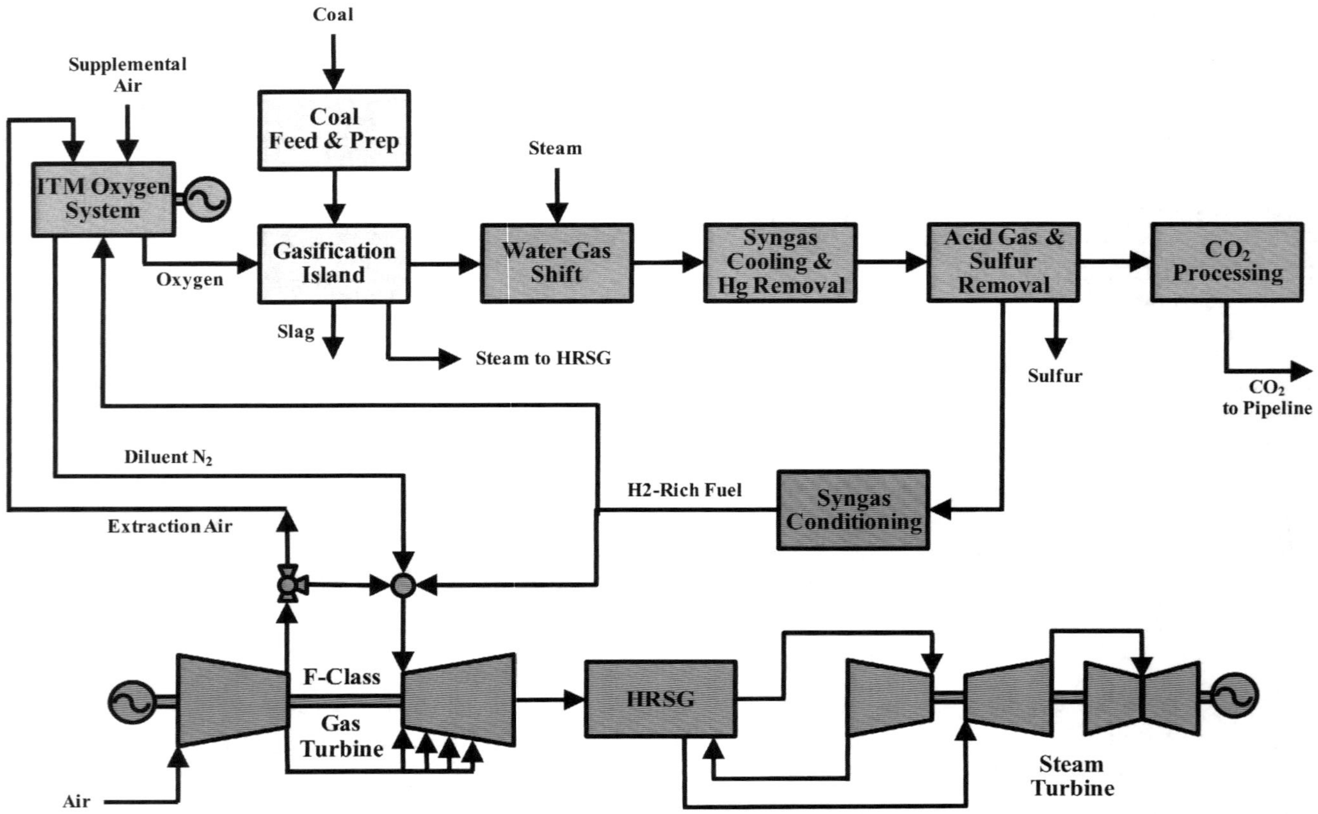

Figure 10. Block flow diagram of the IGCC power generation facility with carbon capture, showing oxygen supply by ITM Oxygen.

After several iterations of design optimization, the F-class IGCC/CCS study focused on comparing results of a single cryogenic ASU case with a single ITM Oxygen case (with sensitivity on the number of gasifiers and gasifier sizing used in the ITM Oxygen case).

- The cryogenic ASU case provided 95 mol% oxygen to two GE 1800 ft3 gasifiers, resulting in essentially an exact fit with expected gasifier capacities to generate the required hydrogen-rich fuel for the F-class gas turbine. The cryogenic ASU design basis included fifty percent gas turbine integration, with 50% of the feed air to the cryogenic ASU's provided by air extraction from the gas turbine (equivalent to 13.5% of the inlet air flow to the gas turbine) and the remaining cryogenic ASU feed air provided by supplemental air compression associated with the cryogenic ASU's. Nitrogen from the cryogenic ASU's was premixed with the hydrogen-rich fuel to the gas turbine to provide diluent for combustion control, NOx mitigation, and to maximize power from the turbine. 87% CO_2 capture was achieved.

- The ITM Oxygen case provided 99.5 mol% oxygen to the GE gasifiers, and approximately 25% additional gasifier and syngas processing capacity (beyond the cryogenic ASU case) was added to produce additional hydrogen-rich fuel for use in the single stage, direct fired ITM Oxygen cycle (e.g. see Figure 7 above). The design basis for ITM Oxygen technology included 13.5% air extraction from the gas turbine (similar to the cryogenic ASU case), which resulted in significantly reduced 22.5% feed air integration to the ITM Oxygen system (relative to the cryogenic ASU at 50% feed air integration). The majority of the feed air requirement to the ITM Oxygen units was provided by supplemental air compression associated with the ITM Oxygen system. A portion of the nitrogen-rich non-permeate from the ITM Oxygen system (with less than 2 mol% oxygen) was generated as diluent and premixed with the hydrogen-rich fuel to the gas turbine to provide for combustion control, NOx mitigation, and to maximize power from the turbine, and additional power was generated within the ITM Oxygen cycle via letdown of the additional hot, high pressure nitrogen-rich non-permeate from the ITM Oxygen units. 87% CO_2 capture was achieved.

It was recognized that 87% CO_2 capture achieved in the IGCC/CCS cases (discussed above) using reasonable design criteria for the syngas generation and purification technologies selected as the study basis is slightly short of U.S. DOE 90% CO_2 capture target. While 90% CO_2 capture was possible in the study cases, such high capture would have required aggressive design in nearly all aspects of gasification and syngas processing, including gasifier design and operation for higher methane conversion, addition of a third shift reactor stage for higher carbon monoxide conversion, aggressive Selexol design and operation for higher CO_2 capture, and an expectation of more frequent catalyst change-outs and higher maintenance costs. Such aggressive design and cost did not appear justified in this particular study, and the design basis was maintained at 87% CO_2 capture.

Complete plant designs and cost estimates were generated for the two Integrated Gasification Combined Cycle (IGCC) Power Plant with Carbon Capture cases described above, and Levelized Cost of Electricity (LCOE) analysis was performed using methods and pricing similar to that used in the Oxy-coal Combustion cases discussed above. In doing so, it was recognized that the analysis was appreciably dependent on gasifier sizing and utilization. Discussed above, the cryogenic ASU IGCC/CCS case was essentially an exact fit with two GE 1800 ft3 gasifiers. The ITM Oxygen IGCC/CCS plant design required approximately 25% additional hydrogen-rich fuel for use in the ITM Oxygen cycle, resulting in a capacity mismatch using standard GE gasifiers. To ensure the results of the study maintained intended focus on comparing oxygen supply technologies, three design and capital cost sensitivity cases were performed on gasifier train size and cost, including case 2A with two large gasifier trains at 100% utilization, case 2B with three under-utilized standard GE 1800 ft3 gasifier trains, and case 2C with three small gasifier trains at 100% utilization. Case 2A would appear the best comparison against the cryogenic ASU IGCC/CCS case, as it maintains the same number of gasifier trains at 100% utilization.

Summary of the Integrated Gasification Combined Cycle (IGCC) Power Generation with Carbon Capture study results is presented in Table 2. The ITM Oxygen case (case IGCC-2A) showed 1.8 %pts of plant efficiency advantage, 12% lower specific capital cost, and 10% reduction in levelized cost of electricity relative to the cryogenic ASU case (case IGCC-1). The ITM Oxygen technology case also showed significant increase in net power output, raising the facility power production levels back to the nominal IGCC rating without carbon capture.

Table 2. Integrated Gasification Combined Cycle (IGCC) Power Generation with Carbon Capture study results.

Oxygen Technology	Cryogenic ASU	ITM Oxygen		
Case	IGCC-1	IGCC-2A	IGCC-2B	IGCC-2C
Gas Turbine Power (MW)	464	464	same as 2A	same as 2A
Steam Turbine Power (MW)	301	325	same as 2A	same as 2A
ITM Oxygen Net Power (MW)	0	112	same as 2A	same as 2A
Total Gross Power (MW)	765	901	same as 2A	same as 2A
Total Auxiliary Load (MW)	206	130	same as 2A	same as 2A
Net Plant Power Output (MW)	559	771	same as 2A	same as 2A
Net Plant Power Output (MW, Δ%)	base	+ 38%	same as 2A	same as 2A
Net Plant Efficiency, HHV (Δ%pts)	base	+ 1.8 %pts	same as 2A	same as 2A
CO2 Captured (kg CO_2/MWH-net)	881	828	same as 2A	same as 2A
CO2 Emitted (kg CO_2/MWH-net)	110	105	same as 2A	same as 2A
CO2 Capture Percentage	87%	87%	same as 2A	same as 2A
Gasifier Trains	2 x 1800 ft3	2 x 2250 ft3	3 x 1800 ft3	3 x 1500 ft3
Gasifier Utilization	100%	100%	83%	100%
Total Plant Cost ($, Δ%)	base	+ 21%	+ 30%	+ 25%
Specific TPC ($/KW-net, Δ%)	base	- 12%	- 5%	- 9%
Levelized Cost of Electricity ($/MWH, Δ%)	base	- 10%	- 5%	- 8%

iii. Summary of Power Generation with Carbon Capture Results

A simplified comparison across both the Oxy-combustion and IGCC studies (discussed above) is presented in Table 3.

- The ITM Oxygen-based power generation facilities show consistent advantage over comparable cryogenic ASU-based power generation facilities. Advantages appear in higher net plant power output, higher net plant efficiency, lower specific total plant cost, and lower levelized cost of electricity.
- The Oxy-coal Combustion facility using ITM Oxygen technology to achieve 73% CO_2 capture shows the highest efficiency, lowest specific capital cost, and lowest levelized cost of electricity.
- The IGCC facility using ITM Oxygen technology to achieve 87% CO_2 capture shows the second highest efficiency, second lowest specific capital cost, and second lowest levelized cost of electricity.

Table 3. Comparison of Power Generation with Carbon Capture study results: Oxy-coal Combustion vs. Integrated Gasification Combined Cycle (IGCC).

	Oxy-Combustion 90% CO2 Capture		Oxy-Combustion 73% CO2 Capture		IGCC 87% CO2 Capture	
Oxygen Technology	Cryo ASU	ITM Oxygen	Cryo ASU	ITM Oxygen	Cryo ASU	ITM Oxygen
Gas Turbine Power (MW)	0	0	0	0	464	464
Steam Turbine Power (MW)	800	800	800	800	301	325
ITM Oxygen Net Power (MW)	0	177	0	370	0	112
Total Gross Power (MW)	800	977	800	1170	765	901
Total Auxiliary Load (MW)	267	150	247	132	206	130
Net Plant Power Output (MW)	533	827	553	1039	559	771
Net Plant Power Output (MW, Δ%)	base 1	+ 55%	base 2	+ 88%	base3	+ 38%
Net Plant Efficiency, HHV (Δ%pts)	base 1	+ 2.5 %pts	base 2	+ 7.6 %pts	base 3	+ 1.8 %pts
CO2 Captured (kg CO2/MWH-net)	1,035	792	783	525	881	828
CO2 Emitted (kg CO2/MWH-net)	78	89	289	197	110	105
Total Plant Cost ($, Δ%)	base 1	+ 27%	base 2	+ 25%	base 3	+ 21%
Specific TPC ($/KW-net, Δ%)	base 1	- 18%	base 2	- 34%	base 3	- 12%
Levelized Cost of Electricity ($/MWH, Δ%)	base 1	- 5%	base 2	- 20%	base 3	- 10%

Detailed engineering study of a nominal 1,000 MWe (gross) coal-based IGCC facility using advanced H-class gas turbine technology with carbon capture to compare the supply of oxygen from ITM Oxygen with traditional cryogenic air separation units (ASU) is ongoing.

Future work

The ITM Oxygen development program is currently focused on completing construction of an Intermediate-Scale Test Unit (ISTU) for operation and testing in 2013. The ISTU is located at an Air Products commercial air separation plant in Convent, Louisiana, and is sized for a nominal 100 TPD of oxygen production. The ISTU will provide design, operation, and scale-up data necessary for design of similarly-sized commercial units, as well as for design of a nominal 2000 TPD test facility required for scale-up to clean energy plants by the middle of the decade. In this development scenario, world-scale ITM Oxygen plants for IGCC and other large clean energy applications would be available by around 2017.

The primary goal of the ISTU is to test large arrays of ITM Oxygen modules housed in a common pressure vessel, while operating coupled to dedicated turbo-machinery configured for power generation. The testing will enable further scale-up of the vessel system and verify process control schemes required to maintain membrane module integrity and turbo-machinery performance. In addition, performance verification will be accomplished for scaled-up supporting systems, such as heat exchange, contaminant control, and oxygen vacuum systems.

The ISTU process configuration is shown in Figure 11 as a simplified block flow diagram. The arrangement comprises two separate sub-systems: one making power, consisting of the air compression train, low-temperature recuperator, combustor, and hot gas expander; and the other making oxygen, consisting of a high-temperature recuperative heat exchanger, an indirectly-fired heater, the ITM vessel, and an oxygen cooling and compression train. As such, the power production process can be operated somewhat independently of the ITM process, thus isolating each system from the other, especially during startup and other transients, and thereby increasing the overall reliability of the process. The natural demarcation between these two sub-systems occurs on the cool side of the high-temperature recuperator, denoted in Figure 11 by the red dashed line, which allows selection of standard control system hardware to accomplish the desired compartmentalization. Reference 10 contains more detailed discussion of the ISTU configuration and future test plans.

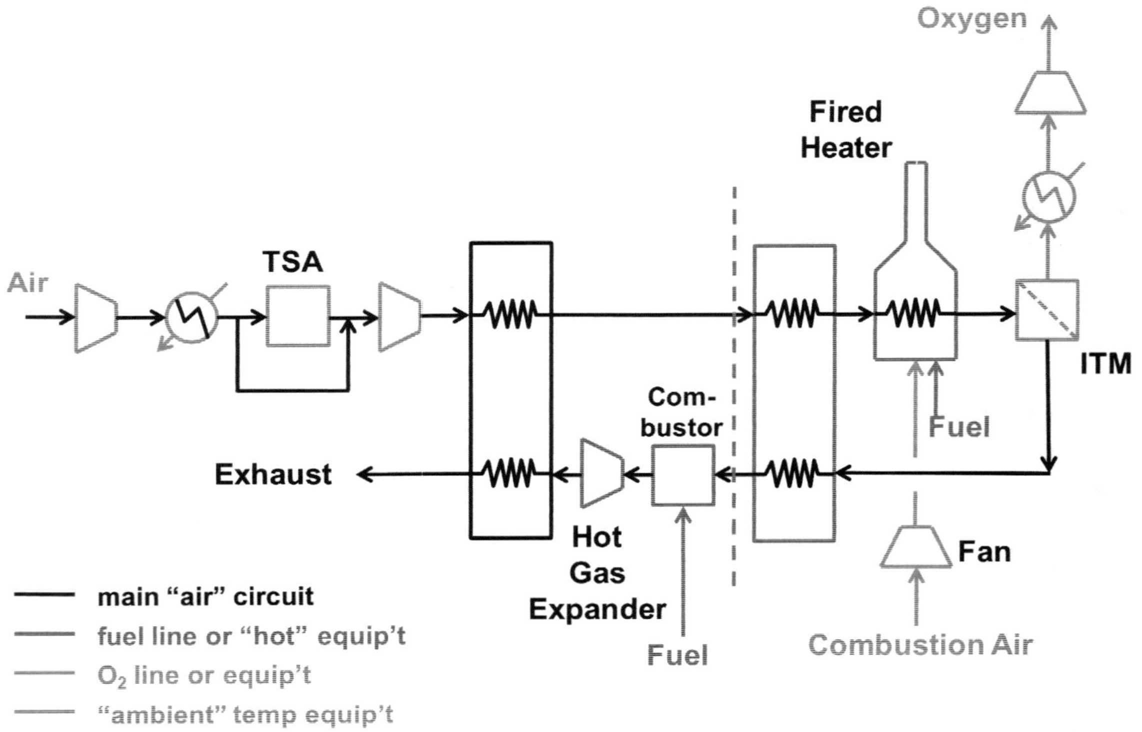

Figure 11. Simplified Block Flow Diagram of the 100 TPD Intermediate-Scale Test Unit (ISTU).

The ITM Oxygen development program has also been focused on ceramic processing development to enable scale-up of ceramic membrane modules to commercial scale. In addition, Air Products, working through its subcontractor Ceramatec, Inc., has been developing methods to manufacture membranes which are scale-able to large-scale fabrication operations characteristic of those in the ceramic processing industry. During Phases 1-3 of the U.S. DOE Cooperative Agreement, Air Products through its subcontractor Ceramatec produced membrane modules with a prototype development line that has supplied small scale test apparatus and the SEP, and which will supply the ISTU with test articles. All of the main processing steps have been established through extensive testing and monitoring of process yields. The prototype development line has provided a good design basis for further scale-up under Phase 5.

With further funding under the American Recovery and Reinvestment Act (ARRA) directed through the U.S. DOE under Phase 5 of the program, Air Products, with subcontractor Ceramatec, is carrying out additional scale-up of ceramic manufacturing in a dedicated large-scale manufacturing facility (the "CerFab") that will be sized to produce enough modules for the future 2000 TPD ITM Oxygen test facility mentioned above. Air Products will design, construct, and commission the CerFab and will assess the operating capabilities of the CerFab during a short-term module-production campaign in 2014. In addition, advanced materials and production techniques will be developed to support the design and operation of the CerFab.

Process design tasks will be carried out to bring definition to early applications and the future 2000 TPD test facility.

Since the Phase 5 DOE program award in late 2010, Air Products has secured a manufacturing site near Salt Lake City in which to begin construction (see Figure 12). Detailed engineering, process design and layout, equipment selection, and environmental permitting were completed in 2012. Facility modifications and equipment procurement is well underway. Detailed discussion of manufacturing steps developed for fabrication of ITM Oxygen wafers and CerFab configuration were published previously in reference 10.

Figure 12. Ceramic manufacturing facility (the "CerFab") will be located in Tooele, Utah, near Salt Lake City. Modifications for this facility are underway; process layout (shown on right) is complete.

Commercial Deployment Vision

Discussed above, the Sub-scale Engineering Prototype (SEP) started up at Sparrows Point, Maryland in 2006 and has achieved its performance goals over multiple operating runs. Expected values of purity and flux have been measured at commercial module scale, and more recent work at the SEP includes performance testing of components and design concepts planned for next level scale-up of ITM Oxygen technology. Current program activities are focused on construction of the next level 100 TPD Intermediate Scale Test Unit (ISTU) at Convent, Louisiana, which is scheduled for operation in 2013. Parallel focus is also directed toward construction of a dedicated large-scale ceramic manufacturing facility (the "CerFab") sized to produce modules for a future 2000 TPD ITM Oxygen test facility; initial module-production is scheduled mid-year 2014. Success in both the ISTU operating experience and the CerFab module production will be critical milestones toward the next scale test facility and commercialization of ITM Oxygen. A 2000 TPD ITM Oxygen test facility and potential commercialization of ITM Oxygen around 100's TPD oxygen scale for early industrial applications are envisaged toward the middle of this decade. Scale-up and deployment of ITM Oxygen into large-scale future energy applications is possible by the end of the decade as shown graphically in Figure 13.

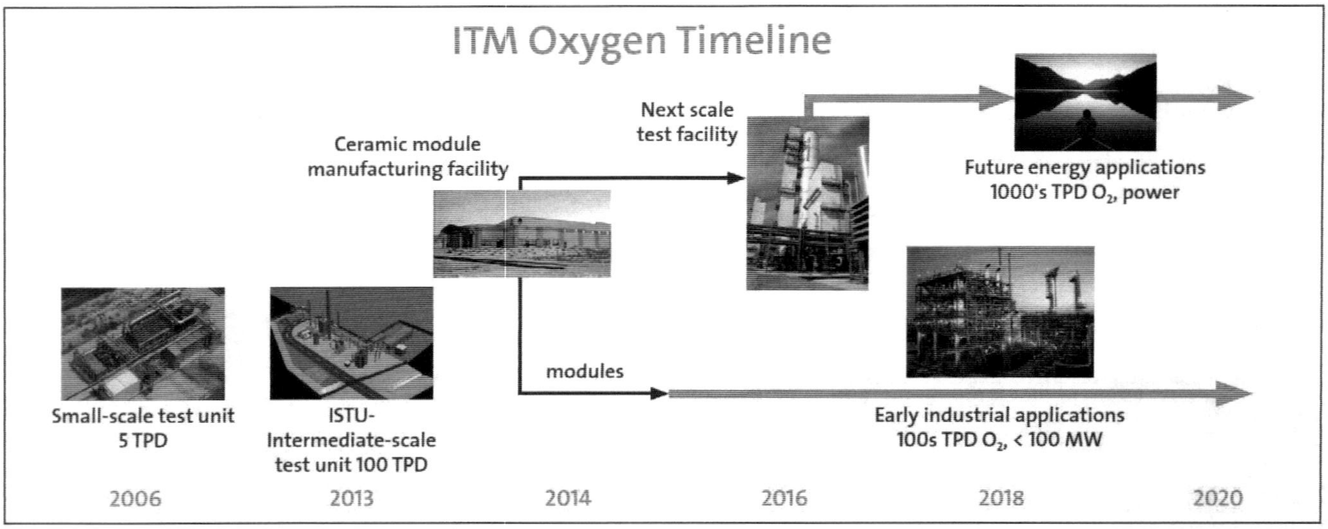

Figure 13. Envisaged scale-up and commercial deployment timeline for ITM Oxygen.

Conclusions

Excellent results have been achieved in the prototype ITM Oxygen unit, confirming expected membrane performance in terms of flux and purity achieved. Multiple thermal cycles have been performed maintaining target product purity, full size modules were tested and robust operation has been demonstrated. These experimental results support and confirm the excellent economic benefits originally projected for the ITM Oxygen technology. Furthermore, as reported here for the first time, process arrangements have been developed and detailed engineering studies have confirmed the excellent economic benefits of ITM Oxygen technology toward large-scale IGCC and Oxy-combustion power generation with reduced CO_2-emissions.

Construction is nearing completion toward the next step in the scale-up of ITM Oxygen technology to an intermediate-scale ITM test facility (100 TPD) and ceramic manufacturing scale-up is underway to provide a basis for early commercialization deployment and further scale-up to large-scale plants required for coal gasification, oxy-combustion, energy-intensive manufacturing industry, and clean energy applications.

References

[1] D. Wright and R.J. Copeland. Report No. TDA-GRI-90/0303 prepared for the Gas Research Institute, September 1990.

[2] B.C.H. Steele. C. R. Acad. Sci. Paris, t.1, Serie II, 533 (1998).

[3] P.N. Dyer, R.E. Richards, S.L. Russek, and D.M. Taylor. "Ion Transport Membrane Technology for Oxygen Separation and Syngas Production," Solid State Ionics, 134 (2000) 21-33.

[4] H. Jaeger. "ITM Oxygen Offers 15% More Power and Over 10% Better Plant Efficiency," Gas Turbine World 2008 IGCC Reference Guide, pp. 34-38.

[5] V.E. Stein, E. Juwono, and E.P. Demetri. "The Impact of ITM Oxygen on Economics for Coal-based IGCC," Proceedings of the 27th International Technical Conference on Coal Utilization & Fuel Systems, Clearwater, Florida, 4-7 Mar 2002.

[6] L.A. Vratsanos, P.A. Armstrong, R.P. Underwood, V.E. Stein, and E.P. Foster, "ITM Oxygen for the Real World: Pilot Scale Testing and Beyond," Proceedings of International Pittsburgh Coal Conference, Pittsburgh, PA, USA, 29 Sept-2 Oct 2008.

[7] L.A. Vratsanos, P.A. Armstrong, R.P. Underwood, V.E. Stein, and E.P. Foster, "Enabling Clean Energy Production from Coal: ITM Oxygen Development Update," Proceedings of International Pittsburgh Coal Conference, Pittsburgh, PA, USA, 20-23 Sept 2009.

[8] L.L. Anderson, P.A. Armstrong, J.M. Repasky, and V.E. Stein, "Enabling Clean Coal Power Generation: ITM Oxygen Technology," Proceedings of the 28th Annual International Pittsburgh Coal Conference, Pittsburgh, PA, USA, 12-15 Sept 2011.

[9] J.M. Repasky, L.L. Anderson, P.A. Armstrong, E.P. Foster, and V.E. Stein. "ITM Oxygen Development for Advanced Oxygen Supply," Proceeddings of the 2011 Gasification Technologies Council (GTC) Conference, San Francisco, CA, 9-12 Oct 2011.

[10] J.M. Repasky, L.L. Anderson, V.E. Stein, P.A. Armstrong, and E.P. Foster. "ITM Oxygen Technology: Scale-up Toward Clean Energy Applications," Proceedings of the 29th Annual International Pittsburgh Coal Conference, Pittsburgh, PA, 15-18 Oct 2012.

[11] R. Allam, V. White, V. Stein, C. McDonald, N. Ivens, and M. Simmonds. "Revamping Heaters and Boilers to Oxyfiring – Producing Oxygen by ITM Technology," in *Carbon Dioxide Capture for Storage in Deep Geologic Formations – Results from the CO_2 Capture Project: Volume 1*, Chapter 30, Elsevier Ltd., 2005.

[12] R. Steele, P.A. Armstrong, and J.M. Repasky. "Integration of Ion Transport Membrane Technology with Oxy-combustion Power Generation Systems," 2nd International Oxyfuel Combustion Conference, IEAGHG Programme, Yeppoon, Australia, 12-16 Sept 2011.

[13] R.L. Schoff, "The Electric Power Research Institute's Large-Scale Demonstration Program for a Low-Carbon Future," IEA CCT2011 April 9, 2011.

[14] R.L. Schoff, "The Future of IGCC Technology CCPC-EPRI IGCC Roadmap Results," IEA CCT2011 Conference, Zaragoza, Spain, 10 May 2011.

[15] "Advanced IGCC, A Final Phase III Report," Canadian Clean Power Coalition (CCPC) Phase III Results Summary, Appendix A, Prepared by CCPC Technical Committee, June 2011.

[16] J.M. Repasky, V.E. Stein, P.A. Armstrong. "Ceramic and Coal: ITM Oxygen for Power Generation with Reduced CO_2-Emissions," American Institute of Chemical Engineers, Proceedings of the 2009 Spring National Meeting, Tampa, Florida, USA, 26-30 April 2009.

[17] J.M. Repasky. "Process and Method for Heating an Oxygen-Containing Feed Stream to an Ion Transport Membrane System," www.IP.com, IPCOM000183831D, 2 Jun 2009.

[18] "Combustion Heating of an Oxygen-Containing Feed Stream to an Ion Transport Membrane System," www.IP.com, IPCOM000206042D, 12 Apr 2011.
[19] "Heat Exchanger-in-Vessel Design for Use in Processes Comprising an Ion Transport Membrane System," www.IP.com, IPCOM000209179D, 29 July 2011.
[20] S. Nakayama and Y. Noguchi. "Pulverized Coal Combustion in O_2/CO_2 Mixtures on a Power Plant for CO_2 Recovery," *Energy Convers. Mgmt.,* 33(5-8) (1992) 379-386.
[21] K.B. Fogash and V. White. "Oxycoal Combustion: Opportunities and Challenges – Purification of Oxyfuel-Derived CO_2," American Institute of Chemical Engineers, Proceedings of the 2007 Annual Meeting, Salt Lake City, USA, 4-9 Nov 2007.
[22] Engineering Feasibility and Economics of CO_2 Capture on an Existing Coal-Fired Power Plant, Final Report – Volume I. American Electric Power's Conesville Power Plant Unit No. 5 Capture Retrofit Study and Volume II. Bench-Scale Testing and Computational Fluid Dynamics (CFD) Evaluation," submitted by Alstom Power Inc., ABB Lummus Global Inc., and American Electric Power prepared for Ohio Dept. of Development and U.S. DOE/NETL, 29 June 2001.
[23] T.F. Yee, R.S. Srinivasan, and R.M. Thorogood. US patent #5447555.

Acknowledgement

This technology development has been supported in part by the U.S. Department of Energy under Contract No. DE-FC26-98FT40343. The U.S. Government reserves for itself and others acting on its behalf a royalty-free, nonexclusive, irrevocable, worldwide license for Governmental purposes to publish, distribute, translate, duplicate, exhibit and perform this copyrighted paper.

(127b) Carbon Storage and Methane Production Via Permafrost Gas Hydrate Deposits

Garrett C. Fitzgerald, Earth & Environmental Engineering, Columbia University, New York, NY and Marco J. Castaldi, Department of Earth and Environmental Engineering (HKSM), Columbia University in the City of New York, New York City, NY

Carbon Dioxide Storage and Natural Gas Production via Gas Hydrate Exchange Process: Down Hole Combustion Thermal Stimulation

AIChE 2013 Spring Meeting, April 28-May 2. Hyatt, San Antonio, Texas

Garrett C Fitzgerald, Marco J. Castaldi

Department of Earth and Environmental Engineering, Columbia University, 550 W 120 St., Room 926, New York, NY 10027

Methane hydrates present a vast potential for future alternative natural gas resources and have the unique ability to be combined with CO_2 sequestration to produce a carbon neutral CH_4 fuel source. CO_2 forms a stable gas hydrate in the same pressure –temperature environment that naturally occurring CH_4 hydrates exist and can be exploited for an exchange process. This work investigates the exchange process of simulated combustion gasses (CO_2 and N_2) with CH_4 in hydrate bearing porous media systems. Porous sediments are prepared with CH_4 hydrate saturations of 50% pore volume and exposed to pure CO_2, and variousCO_2-N_2 gas mixtures while being subjected to a point heat source for assisted hydrate dissociation. This work will present the gas exchange potential of both pure CO_2 and CO_2 –N_2 blends at various heating rates and exchange conditions. The production of CH_4 is used as a metric to evaluate energy potential of this gas production method, and the extent of gas exchange is used to quantify the degree of carbon intensity of the end use CH_4 fuel. CH_4 production rates via the gas exchange methods are compared to those produced via thermal stimulation alone without the co-injection of any exchange gas as a baseline for benefits/challenges associated with CO_2 storage in CH_4 hydrate bearing porous sediment deposits. Multiple CO_2 injection rates are investigated where flow rates range from the lower limit as the amount of CO_2 that would be released for a given amount of heat energy (CH_4 as a fuel) input to the upper limit where enough CO_2 is injected to replace all available CH_4 hydrate in the formation. The benefits of using N_2 as a co-injectent are relatively unknown in the literature and are proposed to enhance the maximum exchange potential and subsequently produce a greater percentage of the in place CH_4. The results of these experiments should present a better understanding of associated N_2 potential benefits as well as quantify the ideal blend mixtures to achieve the highest gas exchange efficiency.

(127d) Synthesis of High Performance SAPO-34 Zeolite Membrane for CO2/CH4 Separation

Yanfeng Zhang[1], Meng Li[1] and Yuhan Sun[2], (1)Low Carbon Conversion Center, Shanghai Advanced Research Institute, Chinese Academy of Sciences, Shanghai, China, (2)State Key Laboratory of Coal Conversion, Institute of Coal Chemistry, Chinese Academy of Sciences, Taiyuan, China

Synthesis of high performance SAPO-34 zeolite membrane for CO_2/CH_4 separation

***Yanfeng Zhang*[1],Meng Li[1], Yuhan Sun[1]**

[1]Shanghai Advanced Research Institute, Chinese Academy of Sciences, 100 Haike Rd, Pudong New District, Shanghai 201203, P.R China

Introduction

Natural gas usually is contaminated with CO_2, which must be removed to meet pipeline requirement. The presence of CO_2 in natural gas decreases the energy content of the gas, and causes pipeline corrosion in the presence of water. Amine adsorption is a mature technology for CO_2 removal from natural gas, but it suffers from high energy cost for solvent regeneration, especially for natural gas with high CO_2 content. Membrane separation has many advantages, such as low energy cost, low capital investment, flexible size, etc, and polymer membranes have been used for natural gas sweetening at commercial scale. However, polymer membranes cannot treat natural gas with high CO_2 content (>20%), since plasticization effect decreases membrane selectivity significantly. Zeolite membranes have great chemical and thermal stability, and their uniform pore size. Zeolite membranes[1¨C4] have potential for CO_2/CH_4 separation owing to their great thermal, mechanical, and chemical stability, and stability at high CO_2 pressures. SAPO-34, a silicoaluminophosphate with chabazite (CHA) type framework having pore diameter 0.38 nm, is perfect for CO_2/CH_4 separation because CO_2 (0.33 nm kinetic diameter) and CH_4 (0.38 nm) can be separated by molecular sieving mechanism. The selective adsorption of CO_2 in SAPO-34 pores also favours the separation of CO_2. Many factors affect the performance of SAPO-34 membrane, including feed pressure [1,2], Si/Al ratios [3,4], seed size [5], template types [5], membrane thickness [4], cation forms [6], CO_2/CH_4 feed ratio, support properties [7], concentration polarization [8] and template residue [9]. Mei et al prepared ion-exchanged SAPO-34 zeolite membrane and CO_2/CH_4 selectivity was increased by 30%. Here we report a simple one step method of preparing ion-exchanged SAPO-34 membrane by combining template removal and ion-exchange in to one step. Obtained Na, K and Li-SAPO-34 membranes have improved separation performance.

Experimental

Detailed information about synthesis, calcination and gas separation test of SAPO-34 membrane can be found in ref 9. Uncalcined SAPO-34 membrane was soaked in 1wt% $NaNO_3$, KNO_3 or $LiNO_3$ solution for 5 min and dried at 383K for 2 h. Then the SAPO-34 membrane was calcined in vacuum at 673 K for 4 h with 1K/min heating and cooling rates. For comparison, blank SAPO-34 membranes were also calcined in vacuum at the same condition.

Results and discussion

Fig 1 shows the SEM images of SAPO-34 membrane. Typical cubic crystals with perfect intergrowth were found in Fig 1a and membrane thickness is ~ 5 μm (Fig 1b). SAPO-34 crystals with thin plate morphology, as shown in Fig 1c, were used as seed for membrane synthesis. XRD patterns indicate that the obtained crystals and membranes have CHA structure (not shown here).

(a) **(b)**

(c)

Fig 1 SEM images of SAPO-34 membrane and SAPO-34 seed, (a)top view and (b)cross section view and (c)SAPO-34 seed.

Table 1. Separation performance of SAPO-34 membranes (50/50 molar CO_2/CH_4 mixtures, room temperature, 4.0 MPa feed pressure).

Membrane treatment	**CO_2 Permeance mol/(Pa m^2 s)**	**CO_2/CH_4 Selectivity**
1wt% $NaNO_3$ in H_2O	2.1°Á10^{-7}	75
1wt% $LiNO_3$ in Acetone	3.9°Á10^{-7}	70
1 wt% $NaNO_3$ in H_2O	2.6°Á10^{-7}	81
1 wt% KNO_3 in H_2O	2.8°Á10^{-7}	73

Eight membranes calcined directly	3.0°Á10^{-7}	52

Eight SAPO-34 membranes were prepared and calcined directly in vacuum to remove the template. The average CO_2 permeance and CO_2/CH_4 selectivity are 2.8°Á10^{-7} mol/(Pa m^2 s) and 52 respectively (feed pressure 4.6MPa and 50/50 CO_2-CH_4 mixture). Four membranes were soaked in 1wt% salt solution for 5 min and dried at 383K. Then these membranes were calcined at 673K in vacuum to remove template. The melting points of $NaNO_3$, KNO_3 and $LiNO_3$ are 581, 607 and 528K respectively, all lower than calcination temperature of 673K. This indicates that the deposited salt on the membrane surface will be melted during calcination. After template removal, the as prepared H-SAPO-34 membrane will be exchanged to Na, K and Li-SAPO-34 membranes. All four membranes exhibits 40 to 60% increase in CO_2/CH_4 selectivity, compared with control experiment. Ion-exchange of H-SAPO-34 increases basicity of SAPO-34 crystals, which favours the adsorption of CO_2. Ion-exchange may also reduce pore size of SAPO-34 crystals, which might limit the diffusion of CH_4. This result is consistent with Mei's result [6].

Conclusions

A simple method was developed to prepare ion-exchanged SAPO-34 membranes in one step. By combining template removal and ion-exchange into one step, Na, K and Li-SAPO-34 membranes were prepared. The obtained SAPO-34 membranes have much higher CO_2/CH_4 selectivity, which is the result of increased basicity from ion-exchange. The combination of template removal and ion-exchange eliminate one step and greatly reduce synthesis cost. This method can also be applied to other zeolite membranes.

REFERENCES

[1] S. Li, J.G. Martinek, J.L. Falconer, R.D. Noble and T.Q. Gardner, Ind. Eng. Chem. Res., 44 (2005) 3220.

[2] M. A. Carreon, S. Li, J.L. Falconer and R.D. Noble, J. Am. Chem. Soc., 130 (2008) 5412.

[3] S. Li, J.L. Falconer and R.D. Noble, Microporous Mesoporous Mater., 110 (2008) 310.

[4] S. Li, J. L. Falconer and R. D. Noble, Adv. Mater., 18 (2006) 2601.

[5] M. A. Carreon, S. Li, J.L. Falconer and R.D. Noble, Adv. Mater., 20 (2006) 729.

[6] M. Hong, S. Li, H. F. Funke, J.L. Falconer and R.D. Noble, Microporous Mesoporous Mater., 106 (2007) 140.

[7] S. Li, J.L. Falconer and R.D. Noble, J. Membr.Sci., 241 (2004) 121.

[8] A. M. Avila, H. H. Funke, Y. Zhang, J.L. Falconer and R.D. Noble, J. Membr. Sci., 335 (2009) 32.

[9] Y. Zhang, A. M. Avila, H. H. Funke, J.L. Falconer and R.D. Noble, J. Membr. Sci., 363 (2007) 29.

(133a) The Evolution of Chemicals R&D in the Information Age

Felipe Tavares, Jansley Pascoal and Bruno Maia, Intratec Solutions LLC, Houston, TX

Most strategic decisions regarding product and process development need to be made in the initial phase of a project lifecycle. In today's competitive environment, R&D teams are under constant pressure to produce results faster with fewer resources. Decisions must be made based on limited data obtained during a very early phase.

In the past, the difficulty in the access to quality information and the lack of suitable tools led to spend significant time and effort to produce realistic economic estimates, and get a sense of the opportunity attractiveness in early stages of the project workflow. The great dilemma experienced by R&D teams was: develop an estimate too quickly but inaccurate or develop a detailed estimate but spending too much time and effort.

In recent years, the world is experiencing constant technological advances reflected in the increasing flow of information through the internet and the exponential growth in the processing power of computers. The easy access to huge amount of information and innovative engineering tools have completely re-invented the workflow of processes and project development. The combined use of extensive public information sources, such as patents and articles databases; conceptual process and equipment design tools; rigorous process simulators; and consistent cost information are re-shaping how R&D teams are doing and managing research.

This session will cover how to make the most of all available resources to evaluate new products and process, in a fast and accurate way, during the conception stage. Also, it is disclosed how to define metrics to help R&D group to conduct their projects, either internally or through partnerships with research centers and universities.

Currently, there are research firms taking benefit from this approach, significantly reducing cycle time of their projects and increasing the number of alternatives assessed by their staff. Also, there are already companies relying on sale of services based on these concepts, which would be impossible decades ago.

Two cases are presented as illustrative examples: the acrylic acid production from sugar fermentation and ethylene production from ethanol dehydration. These processes technologies were quickly assessed from few public available information and hypothetical assumptions allied with engineering tools.

(133b) Software for Synthesizing Complete Set of Pareto Efficient Designs of Energy Supply Chains According to Cost, Ecological Footprint, and Emergy

Leisha Vance[1], **Botond Bertok**[2], Istvan Heckl[3], Heriberto Cabezas[1] and Ferenc Friedler[2], (1)Office of Research and Development, U.S. Environmental Protection Agency, Cincinnati, OH, (2)Department Computer Science and Systems Technology, University of Pannonia, Veszprem, Hungary, (3)Department Computer Science & Systems Technology, University of Pannonia, Veszprem, Hungary

Methodology and computer aid are proposed for designing sustainable supply chains in terms of sustainability metrics by utilizing the P-graph framework. The methodology is an outcome of the collaboration between the Office of Research and Development (ORD) of the U.S. EPA and the research group led by the creators of the P-graph framework at the University of Pannonia. The integration of supply chain design and sustainability is the main focus of this collaboration. A recent extension to the P-graph framework provides a mathematically rigorous procedure for synthesizing the complete set of Pareto optimal networks subject to multiple objectives and constraints, which include profitability and sustainability in the proposed methodology. Specifically, to evaluate the sustainability of a given process under construction including its supply chain, sustainability metrics are incorporated into the design procedure.

The proposed methodology and the software are demonstrated with the optimal design of a supply chain for providing heat and electric power to an agricultural region with relatively limited land area where agricultural wastes can potentially be recovered as renewable resources. Possible sources of heat and electricity included electricity from the Hungarian grid, and heat and electricity generated from natural gas, corn, corn silage, grass silage, or wood or some combination of these sources. Each supply chain was ranked according to cost, and assessed environmental impacts using the ecological footprint (representing land use burden), and emergy (representing energy resource burden). Decisively feasible supply chains were found with cost variations of +2% to -17% compared to "business as usual" scenarios, i.e., using only natural gas and electricity from the Hungarian grid. For these supply chains, the sustainability profile as represented by the ecological footprint varied from +8% to -78%, and the emergy results ranged from -54% to -93%. Most importantly, it appeared feasibly possible to design supply chains for heat and electricity generation which were both cheaper and more sustainable than the supply chain currently in use.

(133c) Multiobjective Optimization of Solar Energy Driven Desalination Systems

Eman Tora, Chemical Engineering and Pilot Plant Department, National Research Centre, Giza, Egypt and **Tarek Moustafa**, Chemical Engineering Department, Cairo University, El Giza, Egypt

Solar energy offers a sustainable energy source that can be used as thermal energy and electrical energy. Due to energy resources shortage and cost increase through the world, obvious attention is paid for using solar energy systems especially in the processes that require extensive energy quantities like desalination. Desalination of sea water can be divided into thermal desalination and electrical desalination units whereby solar energy can be used to drive both categories. However solar energy has much low operating cost, it suffers from other defects like dynamic profile availability, expensive initial capital cost, and need to large areas. These are interrelated parameters but conflicting as well. Hence, determining the optimum operating and design variables sound a combination of different conflict variables. Thus multiobjective optimization is introduced to tackle these contradictory parameters and to determine the optimum tradeoffs. The target is to reduce the cost and increase the system performance simultaneously. Different types of solar collectors are considered to provide the energy needed to drive the desalination process. Likewise, different desalination units are tackled. Performance throughout the year is addressed. Case study considering the Egyptian weather condition is given to illustrate the feasibility and applicability of the proposed multiobjective optimization approach.

Numerical Simulation in a Novel Polysilicon CVD reactor

HUANG Zheqing, LIU Chunjiang

Abstract: The paper presents a novel polysilicon CVD reactor that the flow pattern of the gas mixture is plug flow, which is fundamentally different from that of traditional CVD reactor. A three dimensional theoretical model of mass, momentum and heat transfer process have been proposed. The velocity and temperature profile of gas mixture in two different reactors have been simulated by using computational fluid dynamic method. The simulation result of the flow field shows that the flow pattern of the gas mixture was plug flow and the flow short circuit between the entrance and exit can be avoided in the novel reactor. The simulation result of the temperature field shows that the novel reactor temperature field can be controlled. By changing the operating parameters, the size of high temperature region which can lead to the generation of silicon powder can be diminished. Furthermore, the energy analysis for traditional reactor show that the total energy consumption is around 70kWh·kg-1 and the radiation loss is about 60% of the total energy loss. The energy consumption of the novel reactor is lower 8% at least than that of the traditional reactor.

Key words: Numerical Simulation; Polysilicon Product; CVD Reactor; Energy Consumption

1. Introduction

With the depletion of the petroleum resources, solar energy is cleaner and more abundant, which have been developed rapidly. The polysilicon as the main raw material of the solar cell will increase sharply in the following years [1-3]. Many processes to produce polysilicon have been tested and patented, but the most popular process is the improved Siemens process, which shares about 80% of total polysilicon manufactured world-wide [4-6]. In the process, more than 60% of the energy is consumed in the decomposition process[7-8]; this indicates that the polysilicon CVD reactor is the most important equipment during the process. The polysilicon CVD reactor, also called a Siemens reactor, consists of a gas-tight chamber where several high-purity silicon slim rods are heated by an electric current flowing through them. Polysilicon is deposited on the seed rods through the thermal decomposition of trichlorosilane in a hydrogen environment.

Polysilicon CVD reactor as the key equipment attracted a lot of research attention during the past few years. For instance, a detailed radiated heat transfer behavior in a 36 rods polysilicon reactor have been investigated[8], and the effect of enlarging the reactor capacity, the impact of thermal shields and the influence of the wall emissivity were studied. Kang et al[9].have investigated the gas nozzle effect on the deposition of polysilicon by monosilane Siemens reactor. Gum et al[10] have presented a patent in which the energy consumption was reduced by providing a thin layer of gold inside a reactor chamber. Pazzaglia et al[11] have presented a thermal shield in Siemens reactor to block thermal radiation emitted from the heated polysilicon rods. Liu et al[12] have developed a novel polysilicon CVD reactor, and the velocity profile and temperature field was simulated by computational fluid dynamic method.

It can be found that the research works mentioned above mainly devoted to describe the polysilicon deposition process and the energy saving strategies for polysilicon CVD reactor. In this paper, a more energy saving CVD reactor [13, 14] with a new flow pattern was developed and detailed analysis have been carried out using computational fluid dynamic.

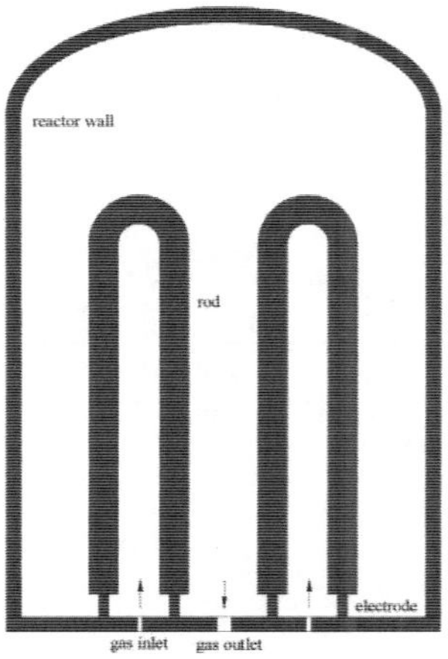

Fig1. Frame structure of traditional polysilicon CVD reactor.

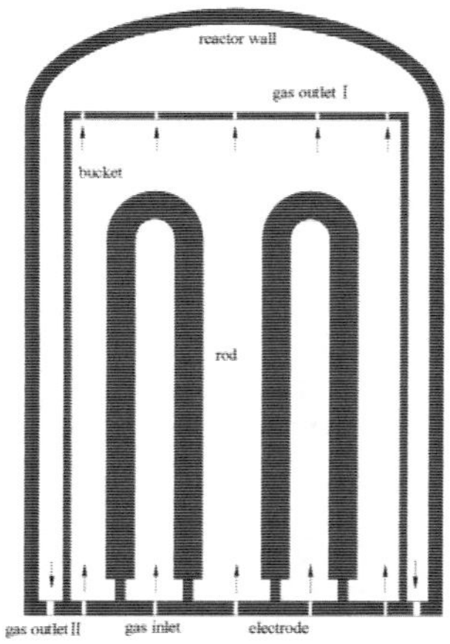

Fig2. Frame structure of novel polysilicon CVD reactor.

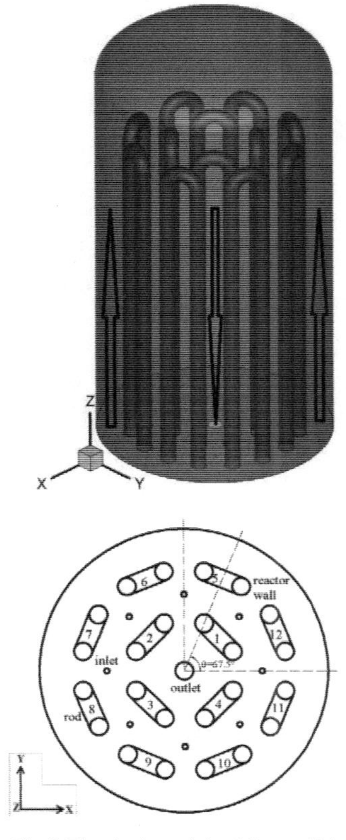

Fig.3 Physical model of the traditional polysilicon CVD reactor.

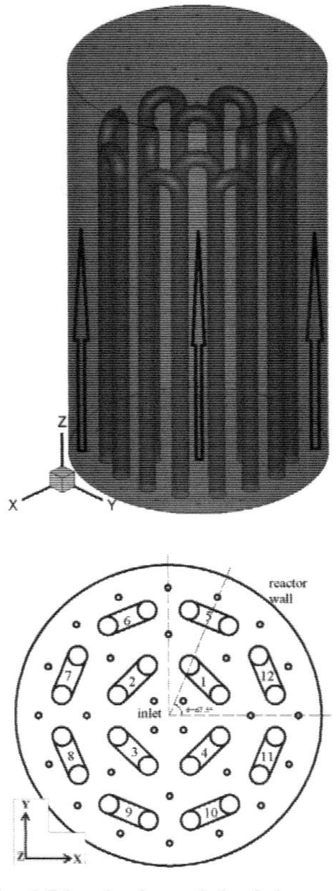

Fig.4 Physical model of the novel polysilicon CVD reactor.

2. Numerical method

2.1 *Physical model*

The frame description of the traditional polysilicon CVD reactor and the novel reactor was given in Figure 1 and Figure 2 respectively. For the traditional reactor, the gases are injected through eight inlets and exhausted through the outlet in the chassis. The difference for the novel reactor is that a bucket is added to change the flow pattern in the reactor, and the gases are exhausted through the outlets in the bucket top. In order to describe the transport phenomena by computational fluid dynamic method, the reactor have been simplified and the physical models with 12 rods have been presented in Figure 3 and Figure 4. Their structure parameters have been presented in Table1.

Table 1 Parameters of the physical model of the CVD reactors

	Traditional	Novel
H/mm	2800	2800
D/mm	1600	1600
h/mm	2300	2300
d_i/mm	20	20
d_o/mm	100	20

2.2 *Governing equations*

The governing equations including continuity, momentum, energy and radiated transfer heat are presented in this section.

1) *Continuity equation*

$$\nabla(\rho \boldsymbol{U})=0 \tag{1}$$

2) *Momentum equations*

$$\nabla\cdot(\rho \boldsymbol{U}\boldsymbol{U}) = -\nabla P + \rho \boldsymbol{g} + \nabla\cdot\left[\nabla \boldsymbol{U} + (\nabla \boldsymbol{U})^{\mathrm{T}}\right] - \nabla\cdot\left(\rho \overline{u'u'}\right) \tag{2}$$

3) *Transport equations for the standard k-ε model*

$$\nabla\cdot(\rho \boldsymbol{U} k) = \nabla\cdot\left[\left(\mu + \frac{\mu_t}{\sigma_k}\right)\nabla k\right] + G_k + G_b - \rho\varepsilon \tag{3}$$

$$\nabla\cdot(\rho\varepsilon \boldsymbol{U}) = \nabla\cdot\left[\left(\mu + \frac{\mu_t}{\sigma_\varepsilon}\right)\nabla\varepsilon\right] + C_{1\varepsilon}\frac{\varepsilon}{k}(G_k + C_{3\varepsilon}G_b) - C_{2\varepsilon}\rho\frac{\varepsilon^2}{k} \tag{4}$$

4) *Energy equation*

$$\nabla\cdot((\rho E + p)\boldsymbol{U}) = \nabla\cdot(\lambda_{eff}\nabla T) + S_h \tag{5}$$

The source term, $S_h = \nabla\cdot(I(\boldsymbol{r},\boldsymbol{s})\boldsymbol{s})$, is the increment of radiation intensity in direction $\boldsymbol{s}$.

5) *Radiative transfer equation*

$$\nabla\cdot(I(\boldsymbol{r},\boldsymbol{s})\boldsymbol{s}) + aI(\boldsymbol{r},\boldsymbol{s}) = a\frac{\sigma T^4}{\pi} \tag{6}$$

In order to calculate the transport properties of gas mixture in the above equations, the properties of species should be obtained. The data for specific heat, viscosity, and thermal conductivity of relevant gas species as a function of temperature has been given in appendix A[15].

2.3 *Boundary conditions*

Walls: No slip boundary condition for the gas velocity has been used on the reactor walls. The isothermal boundary condition is employed. The reactor wall is made of stainless, and the wall surface is polished, and the value of the emissivity for the reactor wall is taken as 0.2[16]

$$\boldsymbol{U}=0,\quad T=373\text{K} \tag{7}$$

$$\boldsymbol{U}=0,\quad T_{Si}=1373\text{K} \tag{8}$$

Rods: No slip boundary condition for the gas velocity is specified, the temperature is 1373K. The emissivity of the polysilicon rods is considered to be 0.7[17].

Inlets: At the inlets, the inlet mass flow, temperature, and the composition of gas mixture are specified, the pressure is 6atm. The radiation absorption coefficient of gas mixture is 0.05[18].

$$Q=0\sim1400\text{m}^3\cdot\text{h}^{-1},\quad T_{inlet}=300\sim600\text{K},\quad Y_{SiHCl_3}=0.25 \tag{9}$$

Outlets: At the outlets, the flow is assumed to be fully developed.

$$\nabla\boldsymbol{U}=0,\quad \nabla T=0 \tag{10}$$

2.4 *Simulation Scheme and Grid strategy*

The polysilicon deposition process in the CVD reactor is non-steady-state, so a series quasi-steady state simulations corresponding to different rod size have been carried out. The physical models and their mesh have been established by using the Gambit. The same grid strategy was adopted in different physical models; the grid with 0.01m was adopted on the polysilicon rods surface; the boundary layer mesh with 0.01m adjacent to the polysilicon rods was adopted; and the tetrahedral mesh with 0.02m was applied in the whole zone. The Fluent software package is used to calculate the governing equations, and the SIMPLE method is used for pressure-velocity coupling.

3. Results and Discussion

The velocity profile, temperature field and the energy consumption of the traditional reactor and novel reactor have been analyzed in the following sections. According to the industrial data (the initial diameter of polysilicon rod is 10mm, and finally diameter is 125mm after 120h reaction), the growth rate of 8μm·min-1 (0.133μm·s^{-1}) is taken in this paper. The quasi-steady state models corresponding to the rods diameter with 29.2mm, 48.4mm, 67.6mm, 86.8mm, 106mm, 125.2mm corresponding to the reaction time from 20 hours to 120 hours have been simulated. Firstly the validation of the proposed model is investigated.

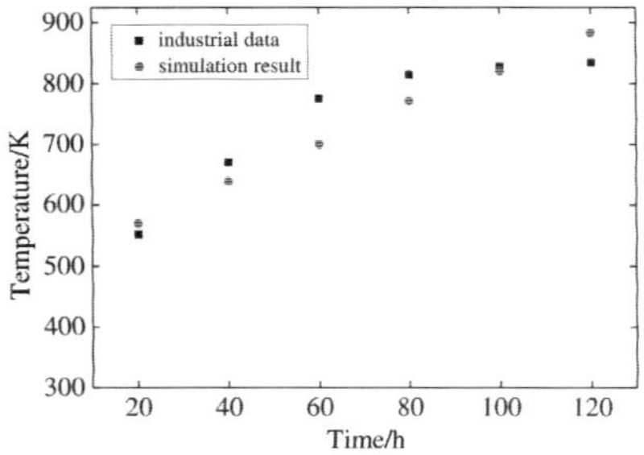

Fig.5 Outlet temperature of the traditional polysilicon CVD reactor.

3.1 *Validation of the CFD model*

In order to verify the validation of the model, the comparison of the outlet temperature between the simulated results and the industrial data obtained by a polysilicon factory in the traditional polysilicon CVD reactor has been done. Fig.5 presents the outlet temperature of the simulated results and the industrial data. It can be found that the deviation of the simulated results and the industrial data is less than 5%. This shows that the proposed model is reliable for describing the behavior of the process.

3.2 *Flow Behavior*

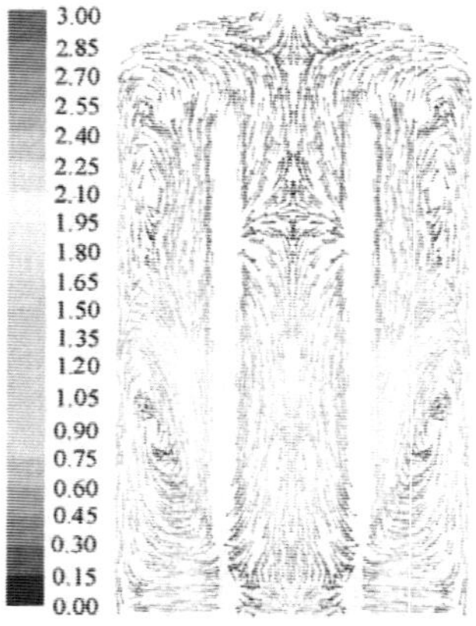

Fig.6 Velocity vector along θ=67.5° plane in traditional reactor/m·s^{-1}

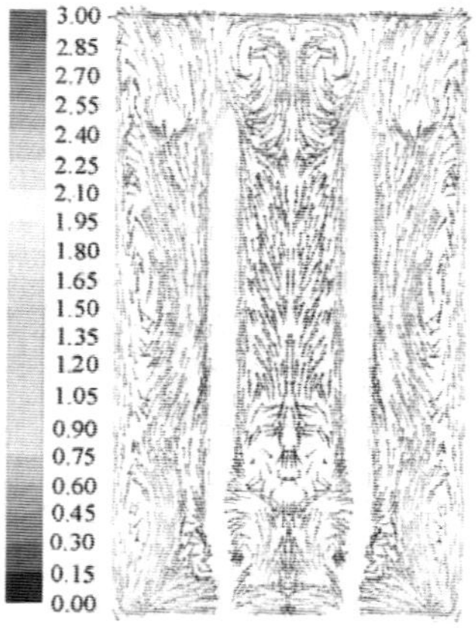

Fig.7 Velocity vector along θ=67.5° plane in novel reactor/m·s^{-1}

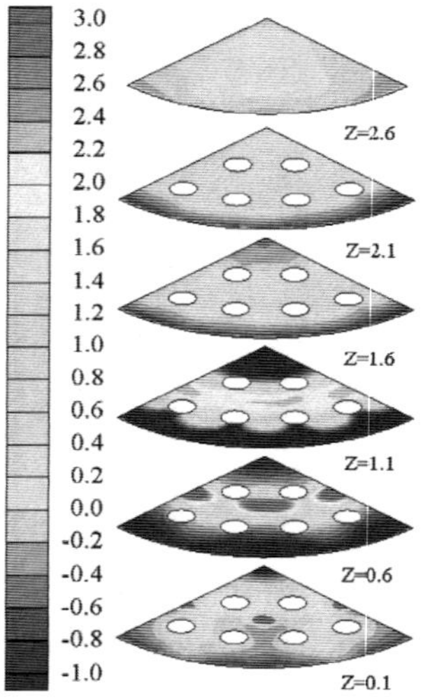

Fig.8 Z velocity contour along different plane in traditional reactor/m·s^{-1}.

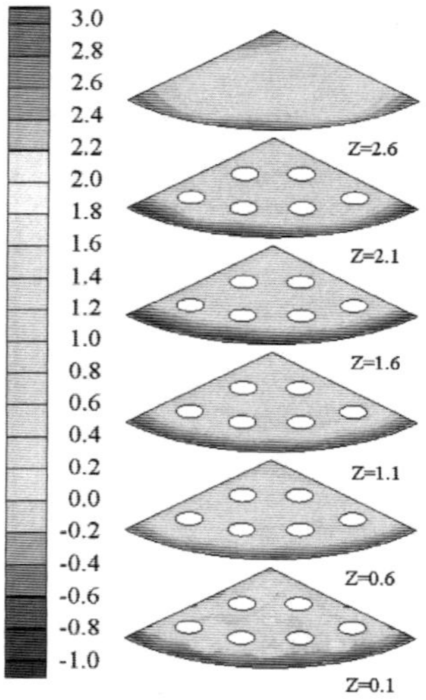

Fig.9 Z velocity contour along different plane in novel reactor/m·s^{-1}.

Fig.6 and Fig.7 give the velocity profile along θ=67.5° plane in different reactors when the rods diameter is 0.106m, and Fig.8 and Fig.9 present the z direction velocity contour along z=0.1, z=0.6, z=1.1, z=1.6, z=2.1, z=2.6 planes in different reactors. It can be found from these figures that the velocity in novel reactor is more uniform than that in traditional reactor. For traditional reactor, a dead zone can be easily formed in the reactor top and the gas mixture go short circuit between the entrance and exit. This problem can be attributed to that the inlets of reactant gases and outlets of exhaust gases in traditional reactor are both opened on the chassis, and the flow pattern of the gas mixture is almost a mixed flow. However, for novel reactor, the problem of the short circuit can be solved and the velocity of the whole zone is almost same, namely the gas flow pattern is almost plug flow (as seen in Fig.7), which is fundamentally different from that of traditional reactor (as seen in Fig.6), so the velocity profile can be easily to control in novel reactor. And to control the velocity profile is very important. This is due to the temperature field distribution greatly affected by the velocity profile.

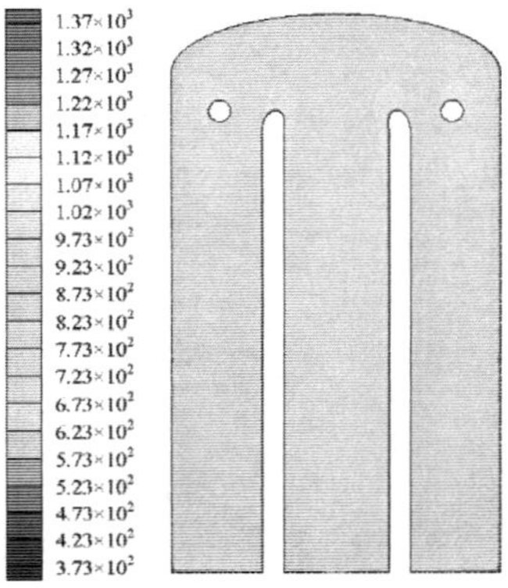

Fig.10 Temperature profile along θ=67.5° plane in traditional reactor/K

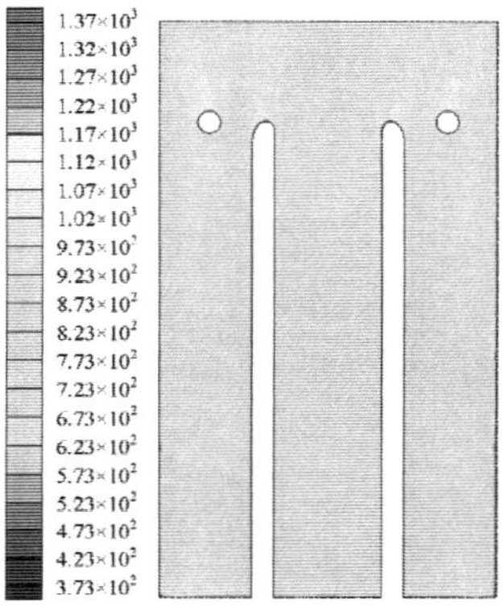

Fig.11 Temperature profile along θ=67.5° plane in novel reactor/K

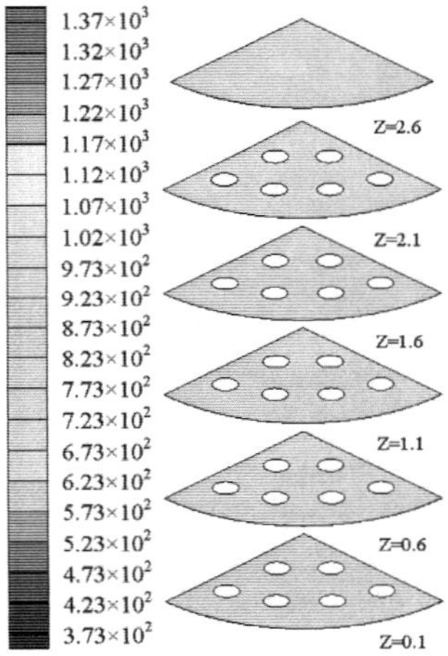

Fig.12 Temperature contour along different plane in traditional reactor/K.

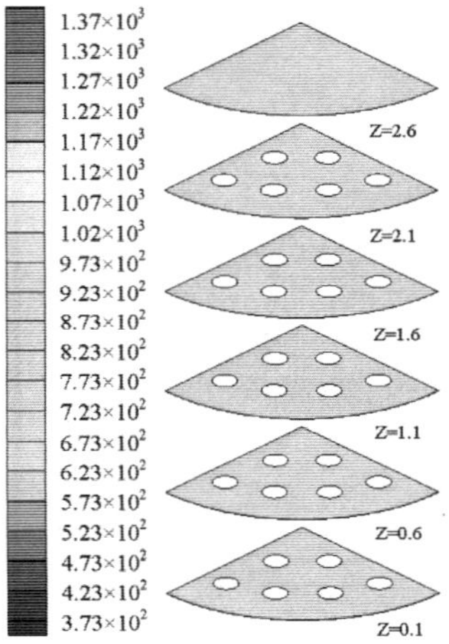

Fig.13 Temperature contour along different plane in novel reactor/K.

3.3 *Heat Transfer Behavior*

Fig.10 and Fig.11 show the temperature field along θ=67.5° plane in different reactors when the rods diameter is 0.106m, and Fig.12 and Fig.13 present the temperature field along z=0.1, z=0.6, z=1.1, z=1.6, z=2.1, z=2.6 planes in two reactors. It is clearly found that the temperature field in traditional reactor is more uniform than that in novel reactor, but the local temperature in traditional reactor is higher than that of novel reactor under the same process conditions. This is due to that the flow pattern in traditional reactor is a mixed flow, thus the temperature field is more uniform. But the simulated result of velocity profile show that the dead zone can be easily formed in the reactor top, so the local temperature is higher. For novel reactor, the flow pattern is almost plug flow, and the dead zone can be eliminated; then the problem of the local high temperature can be solved. It is important for polysilicon CVD reactor to eliminate the dead zone and the high temperature zone, this is due to that the local high temperature will lead to the silicon powder generation, the silicon powder can contaminate the reactor wall during the production process and the cleanness of the wall went dirty, finally the reflectivity of the wall will decrease and the heat loss due to radiation will increase dramatically.

3.4 *Comparison and analysis of energy consumption*

Fig.14 shows the energy loss in traditional reactor and novel reactor throughout the process. It can be found that the energy loss in different reactors decreased with the growth of the polysilicon rods. This is due to the fact that the greater of the polysilicon rods diameter, the more of the polysilicon production per unit time, this means that the average energy loss per unit time and production is lower. When the value of the emissivity in reactor walls are all 0.2, the energy loss is 70kwh·kg^{-1} and 65kwh·kg^{-1} in traditional reactor and novel reactor respectively. And the radiation loss is about 60% of the total energy loss in traditional reactor. The simulated result in traditional CVD reactor is consistent with the industrial data. However, when the value of the emissivity is changed from 0.2 to 0.05[19], for example, the reactor wall and bucket wall were platinum plating (ε=0.05), the energy consumption decrease from 65 kWh·kg^{-1} to 46 kWh·kg^{-1}, which means at least 35% of the energy loss can be saved. This demonstrates the fact that if the cleanness of reactor wall in traditional reactor is contaminated by the silicon powder, the energy loss will increase dramatically.

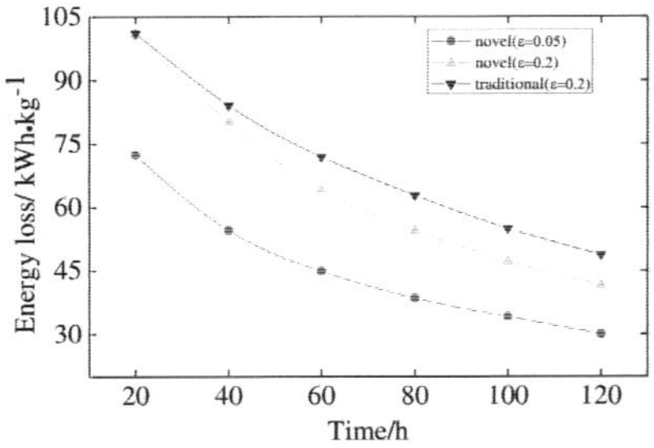

Fig.14 Energy loss for different polysilicon CVD reactors

4. Conclusion

A novel CVD reactor with more energy saving has been developed and a three dimensional theoretical model of mass, momentum and heat transfer process have been proposed. The velocity and temperature profile of gas mixture in two different reactors have been simulated by using computational fluid dynamic method. The simulation result of the flow field shows that the flow pattern of the gas mixture was plug flow and the problem of the flow short circuit between the entrance and exit can be solved in the novel reactor. The simulation result of the temperature field shows that the novel reactor temperature field can be controlled. Due to the temperature field in the novel reactor can be controlled; the high temperature zone in traditional reactor can be eliminated, and the silicon powder generation can be avoided, thus the problem of the inner reactor wall probably contaminated by silicon powder can be solved, and the cleanness of the wall can be remained unchanged. By keeping the cleanness of the inner wall, power loss of the reactor due to radiation can be greatly reduced. The simulation result shows that the total energy consumption is around 70kWh·kg-1 and the radiation loss is about 60% of the total energy loss. The energy consumption of the novel reactor is lower 8% at least than that of the traditional reactor.

Appendix A

The gas density obeys the ideal gas law and the operation pressure is 6 atm;

$$\rho = \frac{PM_{av}}{RT} \tag{A-1}$$

The specific heat of the gas mixture is calculated using the following relationship[20]:

Table2 Heat capacity as a function of temperature

Gases	$J \cdot kg^{-1} \cdot K^{-1}$
Hydrogen	C_p=13749+2.97T-3.84e-3T^2+2.09e-6T^3
Hydrogen chloride	C_p=817.95-0.149T+2.81e-4T^2-8.67e-8T^3
Trichlorosilane	C_p=287.3+1.21T-1.12e-3T^2+3.65e-7T^3
Dichlorosilane	C_p=220.8+1.65T-1.29e-3T^2+3.68e-7T^3
Tetrachloride	C_p=180.1+1.88T-3e-3T^2+2.08e-6T^3-5.27e-10T^3

$$C_p = Y_{SiHCl_3} C_{p,SiHCl_3} + Y_{H_2} C_{p,H_2} \tag{A-2}$$

The viscosity of the gas mixture is calculated using the following relationship[20]:

$$\mu_m = \sum_i^n \frac{y_i \mu_i}{\sum_j^n y_j \phi_{ij}} \tag{A-3}$$

$$\phi_{ij} = \frac{\left[1 + \left(\frac{\mu_i}{\mu_j}\right)^{1/2} \left(\frac{M_j}{M_i}\right)^{1/4}\right]^2}{\left[8\left(1 + \frac{M_i}{M_j}\right)\right]^{1/2}} \tag{A-4}$$

Table3 Viscosity as a function of temperature

Gases	Pa·S
Hydrogen	μ=3.08e-6+2.04e-8T-2.88e-12T^2
Hydrogen chloride	μ=-5.97e-7+5.47e-8T-1.04e-11T^2
Trichlorosilane	μ=4.88E-7+3.78e-8T-3e-12T^2
Dichlorosilane	μ=-1.91e-6+4.8e-8T-1.27e-11T^2
Tetrachloride	μ=2.54e-7+3.77e-8T-1.41e-11T^2

The thermal conductivity of the gas mixture is calculated using the following relationship[20]:

$$\lambda_m = \sum_i^n \frac{y_i \lambda_i}{\sum_j^n y_j \phi_{ij}} \tag{A-5}$$

Table4 Thermal conductivity as a function of temperature

Gas	$W \cdot m^{-1} \cdot K^{-1}$
Hydrogen	λ=0.03754+4.64e-4T-6.72e-8T^2
Hydrogen chloride	λ=1.47e-3+4.51e-5T-4.4e-10T^2
Trichlorosilane	λ=-6.78E-3+5.1e-5T-8.74e-9T^2
Dichlorosilane	λ=3.17e-3+1.08e-5T-4.33e-8T^2
Tetrachloride	λ=8.75e-3-1.59e-5T+5.22e-8T^2

References

[1] C. Del Canizo, G. Del Coso, W. C. Sinke. Crystalline silicon solar module technology: Towards the 1€ Per Watt-Peak Goal[J]. *Progress in Photovoltaics: Research and Applications*, 2009, 17(3):199-209

[2] P. D. Maycock. PV Market Update[J]. *Renewable Energy World*, 2003, 6(4):84-101

[3] S. Pizzini. Towards solar grade silicon: Challenges and benefits for low cost photovoltaics[J]. *Solar Energy Materials and Solar Cells*, 2010, 94(9):1528-1533

[4] J. W. Liang. The Production Technology of Electronic Grade Polycrystalline Silicon[J]. *Engineering Science*, 2000, 2(12):34

[5] A. Luque. Handbook of Photovoltaic Science and Engineering[M]. New York: John Wiley&Sons, 2010.

[6] W. O. Filtvedt, M. Javidi, A. Holt, M. C. Melaaen, E. Marstein, H. Tathgar, P. A. Ramachandran Development of fluidized bed reactors for silicon production[J]. *Solar Energy Materials and Solar Cells*, 2010, 94(12):1980-1995

[7] C. Del Canizo, G. Del Coso, A. Luque. Disclosing the polysilicon deposition process // The 25th European Photovoltaic Solar Energy Conference and Exhibition[C]. Valencia, 2010.

[8] G. Del Coso, C. Canizo, A. Luque. Radiative energy loss in a polysilicon CVD reactor[J]. *Solar Energy Materials and Solar Cells*, 2011, 95(4):1042-1049

[9] S. O. Kang, U. J. Lee, H. J. You, K. Park. Gas nozzle effect on the deposition of polysilicon by monosilane Siemens reactor[J]. *International Journal of Photoenergy*, 2012,

[10]J. C. Gum, C. Fero, D. Desrosier. Gold-coated polysilicon reactor system and method[P]: US, 20110159214. 2011-01-12.

[11]G. Pazzaglia, M. Fumagalli, M. Kulkani. Bell Jar for Siemens Reactor Including Thermal Radiation Shield[P]: US, 2011128729. 2011-10-20.

[12]Z. Q. Huang, C. J. Liu, X. G. Yuan. Numerical simulation of flow and heat transfer in a novel polysilicon reduction furnace[J]. *Chemical Industry and Engineering Society of China*, 2013, 64(2):484-489

[13]C. J. Liu, Z. Q. Huang, L. Duan, X.G. Yuan. A 18 rods novel polysilicon reduction furnace arrangement and connection method[P]: CN, 201110190062.X. 2013-01-02.

[14]X. J. Wang, Z. Q. Huang, L. Duan, C. J. Liu. A 42-rods polysilicon reduction furnace and connection mode[P]: CN, 201110173457.9. 2012-12-19.

[15]R. C. Reid, J. M. Prausnitz, B. E. Poling. The Properties of Gases and Liquids[M]. NewYork: McGraw-Hill, 1987.

[16]W. Bauer, H. Oertel, M. Rink. Spectral Emissivities of Bright and Oxidized Metals at High Temperatures // The Fifteenth Symposium on Thermophysical Properties[C]. Boulder, 2003.

[17]T. Sato. Spectral Emissivity of Silicon[J]. *Japanese Journal of Applied Physis*, 1967, 6(3):339-347

[18]J. L. Qian, Z. H. Yu, L. T. Wang. Tube Furnace[M]. Beijing: Hydrocarbon Processing Press, 1987.

[19]W. L. Wolfe, J. Z. George. Infrared Information and Analysis Center[M]. Washington, D.C.: Office of Naval Research, Department of Navy, 1985.

[20]L. Y. Carl. Matheson gas data book[M]. NewYork: McGraw-Hill, 2001.

(133e) Catalytic Transesterification of Waste Cooking Oil to Biofuel and Conversion of Glycerin with Supercritical Methanol

Chee Kai Tan and Maoqi Feng, Div. of Chemistry and Chemical Engineering, Southwest Research Institute, San Antonio, TX

In the conventional transesterification of fats/vegetable oils for biodiesel production, free fatty acids and water always present a problem and result in a low conversion. The object of this study was to investigate the yield of methyl esters treated by acid-free and alkaline-free with and without a heterogeneous catalyst in supercritical methanol (SCM) condition. Glyceride tributyrate was used as a model compound for waste cooking oil to study transesterification with methanol at supercritical condition (239.4^{o}C and 1140 psi). The transesterification reaction was studied with an automatic controlled fixed bed tubular reactor in a continuous mode as shown in Figure 1. The reactor was constructed from a 3/8" OD x 4' long stainless tube with wall thickness of 0.049" (reactor volume = 47.4 mL). This system was continuously used for the biodiesel synthesis from and waste cooking oil and animal fats.

Figure 1. Continuous Process Flow Diagram for biodiesel production with supercritical methanol.

For glyceride tributyrate, the conversion was 100% under SCM condition. For waste cooking oil, the fatty acid methyl esters (FAMEs) yield was >90% at the following reaction conditions: reaction temperature = 300°C and 250°C, reaction pressures = 2000 psi, 2500 psi, and 3000 psi, residence time = 18 minutes, and molar ratio of methanol to oil = 40:1, flow rate = 2.5 mL/min. However, when feed flowrate increased to 7 mL/min. 250 ^{o}C and 2500 psi, the transesterifcation conversion dropped to <86 %. Heterogeneous catalyst, ETS-10, lowered the transestrification reaction to sub-supercritical conditions, the conversion was consistently achievable over 99% for a 24 hours study at a flow rate of 6mL/min (LHSV 4.47 g. min./mL). Other nanoporous solid catalysts were also studied in similar SCM conditions. The catalysts include hydrotalcite, Zeolite-X, Zeolite-Y, sulfate-$ZrSO_2$, ML-3, and ZSM-5. The findings demonstrated that, by a supercritical methanol approach with heterogeneous nanoporous catalysts, vegetable oil and its wastes could be readily used for biodiesel fuel production in a simple, continuous, and automatic process.

Glycerin is a main byproduct in triglyceride transesterification for biodiesel production as shown in Equation 1.

CH_2-OCO-R CH_2-OH

I I

CH-OCO-R + 3R'OH → 3R-COOR" + CH-OH Eq (1)

I Alcohol Alkyl Ester I

CH_2-OCO-R Biodiesel CH_2-OH

Triglyceride (byproduct) Product Glycerol

R = fatty acid chain R' = CH_3 : fatty acid methyl esters R" = CH_2CH_3 : fatty acid ethyl esters

Continued expansion of biodiesel capacity has led to a rapid increase in glycerin supply, the supply of glycerol to the market could affect the need for chemical derivatives. When the transesterification reaction was conducted at 350ºC with SCM, there was no phase separation of the byproduct glycerol and biodiesel; both components were dissolved in the methanol solvent. GC-MS analysis showed that there were some oxygenated low molecular weight compounds, which are miscible with biodiesel. After carefully analyzing these compounds, glycerol had reacted with methanol and produced isobutanol and 3-methoxy-1,2-propandiol at the higher temperature.

At lower temperature (300 ^{o}C) in SCM, spiked glycrol remained unconverted. Heterogeneous catalysts, ETS-10 and Zeolite-X, contain basic surface states had catalyzed the complete conversion of the glycerol bio-based byproduct to useful fuel additives with higher carbon/oxygen ratio (C/O) at sub-supercritical temperature and pressure.

Research on Isobaric Vapor-Liquid Equilibrium for 1-Propanol-Water-Mixed Solvent System

Xiao-hua LI, Zhi-chao XU, Jian-bing JI

(College of Chemical Engineering and Materials Science, Zhejiang University of Technology,

Zhejiang Province Key Laboratory of Biofue, Hangzhou, Zhejiang, China 310032)

Abstract: Isobaric vapor-liquid equilibrium(VLE) data for 1-propanol-water system containing mixed solvent were measured in a novel ebulliometer at 101.3kPa. The experiments showed mixed solvent(ethanediol and postassium acetate) lead to eliminate the azeotropic point of 1-proanol-water system and rise the relative volatility of 1-propanol to water prominently. The relativity of 1-propanol to water increased with more addition of mixed solvent, so mixed solvent can be considered as an extraction solvent for extraction distillation of 1-propanol and water. The suitable mass fraction of mixed solvent was about 60%. The VLE data were correlated by NRTL model, and comparing the correlated results with the experimental ones, the standard deviations of bubble point temperatures and vapor compositions were 0.45K and 0.00872, respectively.

Key words: vapor-liquid equilibrium ,1-Propanol, water , ethanediol , postassium acetate

1 Introduction

1-Propanol is an important chemistry material and organic solvents. It widely used effective solvent for chemical, pharmaceutical and pesticide industry[1]. In fact, 1-propanol and water has azeotrope. The binary mixtures exhibits a minimum boiling temperature azeotrope with 0.40 of 1-propanol molar composition at 87.67℃under atmospheric pressur.VLE data are useful for the study of solution properties of non-ideal mixtures as well as for the design of the separation equipment. It is relatively difficult to separate the 1-propanol from water by traditional rectification. The extractive distillation method suggested to solve this problem by using an additional agent to alter the relative volatility.

Extractive distillation for dehydrating various alcohols has been studied [2-8]. Many solvents were suggested. Duan Zhanting [9], extractive distillation with salt for anhydrous ethanol, Chen Xiaoping[10],separation of azeotrope of methyl ethyl ketone-water system using extractive distillation with salt, the results were very effective. Mixed solvent (80% ethanediol and 20% postassium acetate) was chosen as the entrainer in this paper Then the VLE data at 101.3 kPa for 1-propanol-water-mixed solvent system were measured and correlated.

2 Experimental

2.1 Materials

Water was double distilled. 1-propanol and ethanediol were analytical grade reagents and were used without further purification. Purities of 1-propanol and ethanediol were checked by GC, and the results were>99.5 %.The postassium acetate were dried for 6hr under vacuum at 400K to remove volatile impurities.

2.2 Apparatus and Procedure

VLE data were measured by a novel ebulliometer made of quartz glass featuring vapor-phase sampling and synthetic liquid composition. Experimental device and method of calculating the liquid composition refer the literature see [11]. The modified ebulliometer was shown in Fig 1.The pressure was measured by a U-tube filled with water, with one end connected to a pressure buffer and another to the atmosphere. Atmospheric pressure was measured by a standard barometer(p). The temperature was measured by a standard platinum thermometer and a 6-1/2-digit multimeter(t). The VLE data measurements for the systems were performed in a way of continuous synthesis. At the beginning of the measurement, certain amounts of 1-propanol and mixed solvent were introduced into the ebulliometer to form the first synthesis. The water contents of the samples in the first synthesis were determined by Karl Fischer analysis, added in or taken out of the ebulliometer was weighed by an electronic balance (Mettler-Toledo Bs224s), Vapor-phase compositions were measured by the Karl Fischer method, when the water mole fraction was less than 0.1,or by gas chromatography (GC-17A,shimazu), when the water mole fraction was greater than 0.1.

In order to verify the reliability of the novel ebulliometer, the VLE data for 1-propanol-water system were measured in this ebulliometer at 101.3kPa, the results agreed with literature[12] values，The thermodynamic consistency test was passed for 1-propanol + water binary system by Herrington. It showed that the novel ebulliometer measured VLE data was reliable.

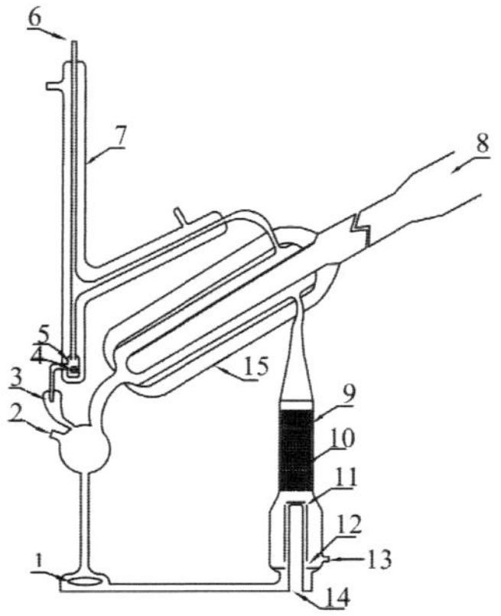

Fig.1 Double circulation ebulliometer

1--Magnetic stirrer functioned as a centrifugal pump；2-Sample injection port；3-Drop counter；4-Vapor phase sampling port；5-Vapor phase sampling chamber；6-Pressure measuring port；7-Vapor condenser；8-thermometric instrument；9-Boiler；10-Column packed by glass coils；11- Holes on the top of the jacket；12-Holes on the lower part of the jacket inside the boiler；13-Bulk material sampling port；14-heating rod；15-Equilibrium cell；

3 Results and discussion

VLE of the system 1-propanol (1) + water (2) + mixed solvent (3) were measured at different mass fractions of mixed solvent, data were measured at w_3=0.400, 0.500, 0.600, 0.700and 0.800. Results were shown in Tables 1. Relative volatility were also calculated. Here x_1' was the liquid-phase mole fraction of component 1, namely, 1-propanol, calculated on mixed solvent free basis. y_1 was the vapor-phase mole fraction of component 1,namely, 1-propanol.

Table 1 Vapor-Liquid Equilibrium Data for the System of 1-Propanol (1) - Water (2) - mixed solvent (3) at101.3kPa

w_3	T/K	x_1'	y_1	a_{12}	w_3	T/K	x_1'	y_1	a_{12}
		w_3≈0.8					w_3≈0.7		
0.8055	393.70	0.8958	0.9418	1.988	0.7016	386.02	0.8964	0.9293	1.651
0.8048	393.48	0.797	0.9013	2.150	0.6982	385.45	0.8001	0.8672	1.864
0.8017	391.67	0.6955	0.8670	2.362	0.7008	384.01	0.6995	0.8320	2.001
0.8015	390.80	0.6036	0.8212	2.691	0.7014	382.64	0.5982	0.7717	2.305
0.8011	390.12	0.5012	0.7480	3.100	0.7023	381.95	0.4975	0.7227	2.690
0.8022	389.21	0.4037	0.7072	3.537	0.7001	380.45	0.3997	0.6508	3.098
0.8003	388.65	0.3021	0.6220	4.120	0.6994	379.60	0.3012	0.5910	3.705
0.8014	388.48	0.1954	0.5394	4.686	0.6989	378.99	0.1998	0.5107	4.297
0.8009	387.29	0.1005	0.4581	5.317	0.6983	378.13	0.1013	0.4406	5.102
		w_3≈0.6					w_3≈0.5		
0.6053	381.73	0.8997	0.9096	1.415	0.5068	378.62	0.8993	0.9018	1.091
0.5994	380.3	0.7975	0.8431	1.531	0.5011	376.78	0.7968	0.8311	1.281
0.6044	379.16	0.6984	0.7998	1.701	0.5002	376.87	0.7013	0.7570	1.394
0.6029	377.84	0.6011	0.7403	1.981	0.5046	375.15	0.5997	0.6906	1.596
0.6012	376.48	0.4997	0.6822	2.419	0.5033	372.60	0.5028	0.6428	2.000
0.6064	375.30	0.3993	0.6248	2.903	0.5037	371.40	0.4016	0.5891	2.401

0.6031	374.47	0.3002	0.5645	3.301	0.4993	370.71	0.2996	0.5293	3.091
0.5996	373.19	0.2029	0.5003	4.095	0.5052	370.58	0.2001	0.4792	3.697
0.6001	372.23	0.1002	0.4300	4.789	0.5002	369.78	0.0991	0.4154	4.688
		w_3≈0.4					w_3≈0.4		
0.4042	375.86	0.8988	0.8849	0.931	0.4032	368.36	0.4027	0.5511	2.094
0.4032	373.79	0.7998	0.8001	0.965	0.4012	367.67	0.2998	0.5087	2.568
0.4056	372.11	0.7012	0.7221	1.091	0.4028	366.65	0.2059	0.4539	3.367
0.4043	370.99	0.5994	0.6582	1.313	0.3988	365.97	0.1002	0.4017	4.257
0.4042	369.61	0.5052	0.6089	1.651					

The experimental VLE data were correlated with the NRTL equation[13]. In the correlations, the binary parameters for water and 1-propanol were taken from the literature 12. For the simplicity of application, the binary parameters were regarded as temperature independent and the values of a_{12},$a_{13}$and a_{23} were set as 0.3. The remaining four binary parameters were fitted by optimization of the following objective function(1).Results for the optimized parameters and deviations in the correlations were summarized in Table 2

$$F=\sum_{j=1}^{m}[(T_{cal}-T_{\exp})_j^2+(y_{1,cal}+y_{1,\exp})_j^2] \quad (1)$$

Table2 Interaction parameters and standard deviations for NRTL

i	1-Propanol	1-Propanol	mixed solvent	Standard deviation	
j	water	mixed solvent	water	$\sigma_T(k)$	σ_y
$g_{ij}-g_{jj}$	1648.8	7245.7	-2430.2		
$g_{ji}-g_{ii}$	7896.7	389.8	4668	0.45	0.0087
a_{ij}	0.3	0.3	0.3		

The effect of the mixed solvent on the phase behavior of the volatile components can be observed in Figure 1. At a given value of x_1', the vapor-phase mole fraction of 1- propanol y_1, increases with the increase of mixed solvent mass fraction. This indicates the enhancement of the relative volatilities of 1- propanol to water. For the mixture of water+1-propanol, the azeotrope was broken with addition of mixed solvent at w_3=0.500.

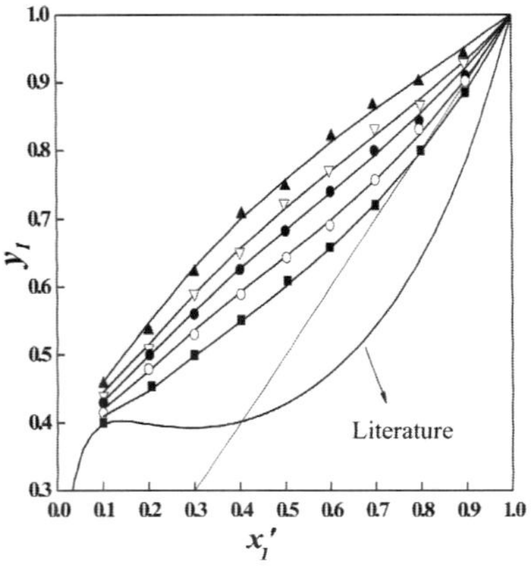

Fig. 2 Vapor-liquid equilibrium curve of 1-propanol-water-mixed solvent system at 101.3KPa

▲$w3$≈0.8; ▽$w3$≈0.7; ●$w3$≈0.6; ○$w3$≈0.5; ■$w3$≈0.4

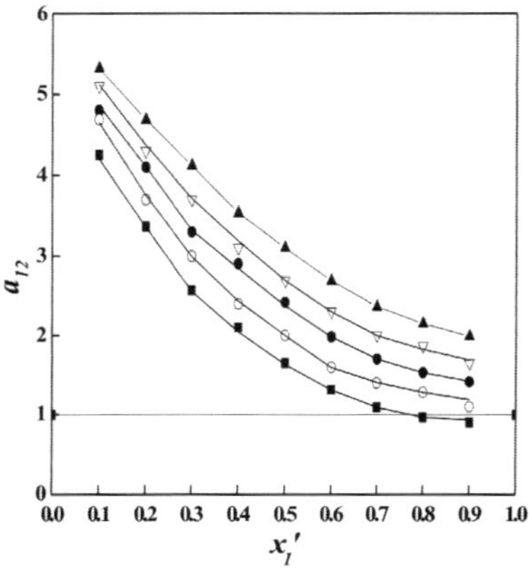

Fig.3 Relative volatility of 1-propanol(1) to water(2) at different mass fraction of mixed solvent

▲$w3\approx0.8$; ▽$w3\approx0.7$; ●$w3\approx0.6$; ○$w3\approx0.5$; ■$w3\approx0.4$

The effect of the mixed solvent on the phase behavior of relative volatility can be observed in Figure 2. At a given value of x'_1, the relative volatility a_{12},increases with the increase of mixed solvent mass fraction. This indicates the enhancement of the relative volatilities of 1- propanol to water. At the same time , when the mixed solvent remain unchanged and 1-propanol mole fraction x'_1 increases, the relative volatility of 1-propanol to water decreases. The effects of vapor-liquid equilibrium can be attributed to its different molecular interaction between these molecules, thus changing the activity coefficient of each component to improve the the relative volatility; That is because the solvent and water are strong polar molecules and they would be priority interaction.Then the activity coefficient of water decreases and 1-propanol increase. That lead to enhancement of the relative volatilities of 1-propanol to water; When the water decreases in system, a few water molecules were priority interaction with the solvent molecules, At the same time, same free solvent molecules will be interaction with 1-propanol.Compared to the former, the relative volatility of 1-propanol to water are decrease. The relative volatility of 1-propanol to water $a_{12}>1$ at $w_3>0.500$. Account into the re-use of solvents and the relative volatility of the separation costs, the option w_3= 0.6~0.7 for the extractive distillation separation of 1-propanol - water azeotrope.

4 Conclusion

1 VLE data for 1-propanol + water + mixed solvent systems were measured by a novel ebulliometer. The vapor composition was obtained by sampling and analysis, while the liquid composition was based on synthetic quantities and was calculated with regard to vapor phase composition, quantity of vapor material, and vapor-to-liquid circulation ratio.

2 The effect of the mixed solvent on the 1-propanol + water azeotrope was obtained by their relative volatility. The experiments showed that mixed solvent(w3>0.5) can eliminate the azeotropic point of 1-proanol-water system and rise the relative volatility of 1-propanol to water prominently. The more mixed solvent adds, the more salting-out effect becomes. The relativity of 1-propanol to water increase with more addition of mixed solvent, so mixed solvent can be considered as an extraction solvent for extraction distillation of 1-propanol and water.

3 The thermodynamic consistency test was passed for 1-propanol + water binary system. The parameters of NRTL models for the ternary system were obtained. The average deviations in temperature and vapor mole fraction were 0.45℃ and 0.00872, respectively.

References

1 LIANG Cheng, Analysis on Development Tendency of Domestic and Global n-propanol Market[J].Economic Analysis of Petroleum and Chemical Industry,2004,10,49-53

2 VERCHER E MUNOZ R. Isobaric vapor-liquid equilibrium data for the ethanol-water-potassium acetate and ethanol-water-(potassium acetate/sodium acetate) systems[J].J Chem Eng Data,1991,36(3):274-277.

3 LIN C L,LEE L S,TSENG H C. Phase equilibria for 1-propanol+water+sodium chloride and +potassium chloride and 2-propanol+water+lithium chloride and +lithium bromide[J].J Chem Eng Data,1993,38(2):306-309.

4 ILLIUTA M C,THYRION F C,LANDEUER O M. Effect of calcium chloride on the isobaric vapor-liquid equilibrium of 1-propanol+water[J].J Chem Eng Data,1996,41(3):402～408.

5 VERCHER E,ROJO F J,ANTONI M A. Isobaric vapor-liquid equilibria for 1-propanol+water+calciu -mnitrate[J].J Chem Eng Data,1999,44(6):1216-1221.

6 WU Wenliang,ZHANG Yaming,Wang Yanru etal. Vapor-liquid Equilibrium Data for System 1-Propanol-water-potassium Acetate[J]. Journal of Chemical Engineering of Chinese Universities，1997,11(4):413-416.

7 BAO Jing,ZHANG Yaming,ZHANG Lei. Vapour-liquid equilibrium for the ethanol-water-mixed solvent systems[J] Journal of Nanjing University of Technology, Vol. 26 No. 5 Sep.2004

8 CHEN Wei miao,ZHANG Ya ming, Recent progress in vapor-liquid equilibrium for alcohol-water-salt systems[J], Journal of Nanjing University of Technology, Vol. 24 No. 6 Sep.2002

9 Duan Zhanting,LEI Liangheng,Zhou Rongqi, Research Advances in Extractive Distillation with Salt[J]. Petrochemical Technology, 1980,9(6):350～353

10 CHEN Xiaoping,ZHANG Zhu, Study on the Separation of Azeotrope of Methyl Ethyl Ketone-water System Using Extractive Distillation with Salt, Journal of Yantai University(Natural Science and Engineering Edition),Vol. 15 No.2 Apr. 2002

11 ZHANG Lianzhong,HAN Jianzhi,DENG Dongshun,etal. Selection of ionic liquids as entrainers for separation of water and 2-propanol[J].Fluid Phase Equilibria,2007,255(2):179-185

12 J GMEHLING,U ONKEN,W ARLT. Vapor-liquid equilibrium data collection[J].Frandfurt dechema chemistry data series,1977-1984,1/6a,122.

13 John M.Prausnitz,Rudiger N.Lichtenthaler,Edmundo Gomes de Azevedo. Molecular Therodynamics of Fluid-Phase Equilibria[M]. BeiJing, Chemical Industry Press,2005,149

14 SANDER B,FRWDENSULUND A,RASMUSSEN P. Calculation of vapour - liquid equalibria in mixed solvent salt systems using an extended UNIQUAC equation[J]. Chem Eng Sci, 1986 ,41(5):1171-1183.

(138a) Synthesis of 2-Acetoxy Propanoic Acid Via Reactive Distillation an Intermediate in Synthesis of Biobased Acrylic Acid From Lactic Acid

Aspi K. Kolah, Lars Peereboom, Carl T. Lira and Dennis J. Miller, Chemical Engineering and Materials Science, Michigan State University, East Lansing, MI

The selective formation of 2-acetoxypropanoic acid (2-APA) by reaction of lactic acid with acetic acid has been carried out in continuous reactive distillation. The reaction takes place in a six-meter tall, 50 mm ID Pyrex column operated at atmospheric pressure and is catalyzed by Amberlyst-15 cation exchange resin supported in Katapak-SP11 structured packing in the middle four meters of the column. Lactic acid solutions in water were used as feedstock in the experimental study which was reacted with glacial acetic acid in the reactive distillation column. High conversions of lactic acid and its oligomers were achieved along with high selectivity to monomeric 2-APA. Experimental studies are complemented by process analysis using AspenPlus simulation software. By implementing a kinetic model and phase equilibrium data generated in our laboratories, we are able to predict the observed pilot-scale column behavior and simulate commercial-scale facilities.

The formation of 2-APA represents the first step of a process to produce acrylic acid from lactic acid. In the second step of the process, 2-APA is cracked under high temperature to form acrylic acid and liberate acetic acid that can be recycled.

Acrylic acid, its esters, and acrylate polymers are critical chemical building blocks used in a wide gamut of applications including superabsorbents, hard plastics, paints, detergents, and adhesives. The global acrylic acid market is estimated at 3.5 million tons and $8 billion annually, with a growth of 4% per year. Acrylic acid synthesized from biobased lactic acid is a drop-in replacement for the petroleum-based product; lactic acid is a mature biobased feed stock with a proven track record of high yield (>95%) from glucose and low cost.

(138b) Revolutionizing the Production of Transportation Fuels and Chemicals

David St. Angelo, Engineering, Joule Unlimited, Bedford, MA

Many technological alternatives to petroleum-based fuels have been explored, and with good reason. Limited energy supplies, pricing volatility and environmental concerns have converged to create one of the greatest challenges of our time. Joule® has an answer that surpasses biomass-derived fuels, channeling the sun's energy into a new, industry-changing category of direct solar fuels. The company's ***Helioculture***™ platform incorporates proprietary, engineered photosynthetic microorganisms to directly produce infrastructure-ready diesel, ethanol and multiple chemicals with no dependence on biomass feedstocks, agricultural land or fresh water. In parallel, Joule has developed a novel ***SolarConverter®*** system to enable the direct, continuous process with productivities greater than biomass-dependent methods, which require numerous energy-intensive steps and downstream processing to achieve an end product. In contrast, using sunlight, non-potable water and waste CO2 from industrial emitters or pipelines, Joule ultimately targets productivities of up to 15,000 gallons of diesel and 25,000 gallons of ethanol per acre annually, at stable costs as low as $50/bble and $1.28/gallon respectively (without subsidies).

(138c) Production and Upgrading Of Renewable Bio-Crude From Large Scale Algae Ponds

Neil Osterwalder, Dan Sajkowski and Ben Saydah, Sapphire Energy Inc., San Diego, CA

Renewable fuels from algae are moving closer to commercialization as Sapphire Energy showed with their Integrated Algae Bio-Refinery (IABR). The IABR in New Mexico was commissioned in 2012 and is producing algae derived bio crude oil on a continuous basis using Sapphire's proprietary conversion and extraction process. This algae bio-crude contains a wide range of hydrocarbons, fatty acids, aromatics, and nitrogen compounds and other heteroatoms. This chemical profile suggests that algal bio-crude can be economically processed by blending with existing refinery streams and co-processing the combined streams as opposed to the more costly development of algal-specific refineries. Different options of co-feeding the algae crude to a refinery can produce a full suite of refinery products from petrochemicals to gasoline, diesel and other refinery products. This study describes the products that can be produced from algae bio-crude and its implementation into existing refining infrastructure.

(138d) Membrane-Assisted Downstream Processing for Biobutanol Purification

Sebastian Heitmann, Daniel Welz, Philip Lutze and Andrzej Górak, Department of Biochemical and Chemical Engineering, Laboratory of Fluid Separations, TU Dortmund University, Dortmund, Germany

One of the main challenges for chemical industry is the change from fossil resources to a sustainable production of chemicals. The change to biobased raw materials and biotechnological processes often leads to product inhibitions of microorganisms and diluted system, resulting in an expensive downstream processing. One bulk chemical that can be produced in a sustainable way is biobutanol. One method of producing biobutanol utilizes the so-called acetone–***n***-butanol–ethanol (ABE) fermentation that can be carried out on basis of e.g. cellulosic and lignocellulosic biomass or agricultural wastes. However, the productivity of this fermentation is still limited, because ***n***-butanol is toxic towards the microorganisms used in ABE fermentation. That is why the concentration of n-butanol in the fermentation broth is often limited to values lower than 2 wt.-percent [1]. For n-butanol recovery mainly distillation processes are used which - as result of the low concentrations - have a high energy demand. About 80 % of the total energy demand of the downstream process is required for the separation of acetone, n-butanol and ethanol from the fermentation broth. Alternatively to distillation the fermentation can be coupled with a pervaporation unit, thus n-butanol can be separated from the broth continuously to prevent the n-butanol concentration from reaching the toxicity limit. Because of the mild operating conditions, pervaporation enables an energy-efficient n-butanol separation and facilitates the recycling of the broth with lower n-butanol content back to the fermenter.

Therefore, the aim of this work was to evaluate if it is more efficient to separate n-butanol by means of pervaporation or distillation or a combination of both. For this purpose extensive experimental studies were performed to characterize the permeation behavior of a commercial available poly(dimethylsiloxane) (PDMS) membrane (Sulzer Pervap™ 4060). These studies provide a wide data basis for the downstream process analysis and are used to identify possible operating windows for the application of a pervaporation. An experimental comparison between the pervaporation of a binary mixture of n-butanol and water, the pervaporation of a synthetic medium and the pervaporation of real fermentation broth was carried out. The synthetic medium contained acetone, n-butanol and ethanol in mass ratios of 3:6:1 together with acetic and butyric acid, which are the solvent precursors. Additionally process parameters like feed composition, pervaporation temperature and permeate pressure were varied. The detailed experimental procedure is described elsewhere [2]. Next to the PDMS membrane, composite membrane materials containing ionic liquids are promising for separation of n-butanol from an aqueous solution. These membrane materials show advantageous permeation properties compared to conventional polymer membranes [2]. Therefore different supported ionic liquid membranes (SILMs) have been tested to evaluate their potential in n-butanol pervaporation.

Based on the experimental results a correlation has been developed describing partial fluxes of acetone, n-butanol, ethanol and water. Fluxes of acetic and butyric acid were neglected. With the help of this correlation modular-based process studies have been carried out using Aspen Custom Modeler™ to find an optimal process configuration for n-butanol separation from the fermentation broth. It was found that pervaporation as stand-alone process as well as distillation is comparatively expensive. Because n-butanol should be almost completely separated from the broth, the concentration at the outlet of the membrane module becomes small, leading to a large required membrane area and a small permeate concentration of n-butanol.

In contrast to the application of distillation and pervaporation as stand-alone processes, a combination of pervaporation and distillation was found to be best suited for efficient n-butanol recovery. A possible membrane-assisted process is shown in Fig. 1. Pervaporation is used for continuous n-butanol removal in order to extend the fermentation cycles. Because the mass fraction of n-butanol $w_{B,Rec}$ in the recycle stream is only slightly smaller compared to the mass fraction of n-butanol in the fermentation $w_{B,Ferm}$, the required membrane area is comparatively small and a higher permeate concentration can be obtained. The complete removal of n-butanol from the broth is then carried out in distillation columns. In this case, the extension of the fermentation periods and the reduction of the stream to be processed in the energy-intensive distillation lead to a reduced energy demand and lowered downstream costs for n-butanol. The permeate stream and the solvent phase from the distillation can be processed further in a conventional distillation sequence. As existing biobutanol production plants mainly rely on distillation sequences, retrofit of a pervaporation thus enables an increase in the biobutanol production efficiency, while existing paid-off plants can still be used.

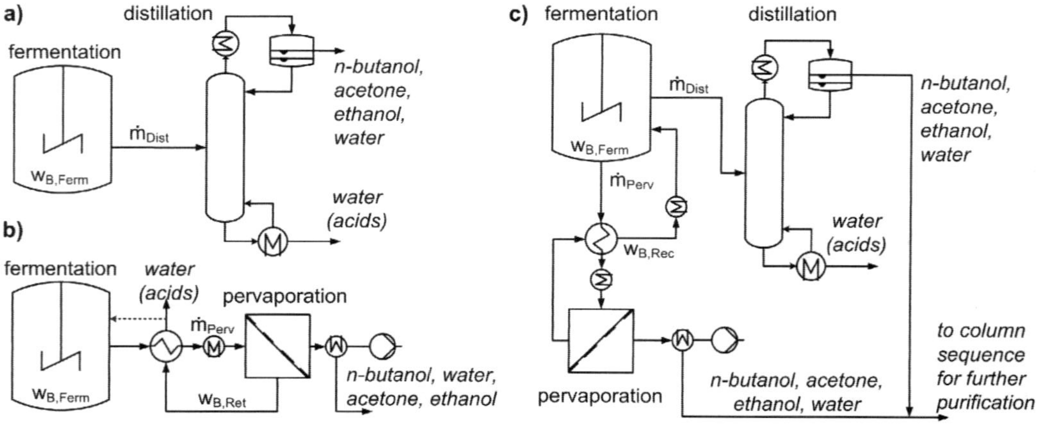

Fig. 1: Solvent removal from fermentation broth by distilliation (a), pervaporation (b) and a membrane-assisted process (c).

Acknowledgement:

The research leading to these results has received funding from the European Union Seventh Framework Programme (FP7/2007-2013) under grant agreement n° 241718 EuroBioRef. We acknowledge Sulzer Chemtech AG and the Laboratory of Biochemical Engineering, TU Dortmund University for providing PDMS membranes and fermentation broth.

[1] T.C. Ezeji, N. Qureshi, H.P. Blaschek, Bioproduction of butanol from biomass: from genes to bioreactors, Curr. Opin. Biotechnol. 3 (2007) 220-227.

[2] S. Heitmann, J. Krings, P. Kreis, A. Lennert, W.R. Pitner, A. Górak, M.M. Schulte, Recovery of n-butanol using ionic liquid-based pervaporation membranes, Sep. Purif. Technol. 97 (2012), 108-114.

Cool Planet Energy Systems and the Production of Carbon Negative Fuels

Mike Rocke, Vice President Marketing
Cool Planet Energy Systems, Inc.
mr@coolplanet.com

Abstract: Cool Planet Energy Systems is a 3-year old start-up producing gasoline blendstock from renewable non-food biomass feedstock. We have devised an economical biomass-to-liquid thermo-chemical conversion process that simultaneously produces liquid fuels and sequesterable biochar useful as a soil amendment. Fuel produced from this process is termed negative-carbon fuel since more carbon dioxide is removed from the atmosphere than released from combustion. The process for producing this fuel relies in part on a unique thermochemical decomposition of biomass that ramps the biomass temperature while the biomass is being subjected to pressure shocks. Co-produced biochar is converted into a soil amendment. We have built a 200,000 gallon per year pilot plant using this process and have demonstrated in laboratory studies that biochar generated using this process can exhibit excellent soil amendment properties. Fuel produced from this pilot plant has been fleet tested in a blend formulation in several standard configuration automobiles. The process is widely tunable in the produced biochar/fuel ratio. We also describe our plans to mass produce trailer-sized modular refinery units using this technology. We are backed by some of the world's largest energy companies and top Silicon Valley venture capitalists. We will discuss the company's trajectory from infancy to present status, and its fit in the bio-economy.

Overview

Cool Planet Energy Systems is a CA based start-up that has developed ground breaking, disruptive technology that economically converts non-food biomass into a sustainable, high-octane gasoline blend stock. Cool Planet's technology is based on twelve awarded, US patents. Hemi cellulosic or lignocellulosic biomass is processed through a mechanical biomass fractionating system that uses pressure and heat to create a series of useful volatile components that then go through proprietary catalytic systems to make high octane fuel. The process also captures the residual carbon to create a byproduct, called biochar, which is an inert, high surface area carbon that enhances soil quality through improved water and fertilizer retention and thereby increasing crop production on marginal or arid farmland. This carbon negative process removes atmospheric carbon to reverse the effects of fossil fuel consumption resulting in up to 150% carbon footprint reduction for every gallon of fuel used.

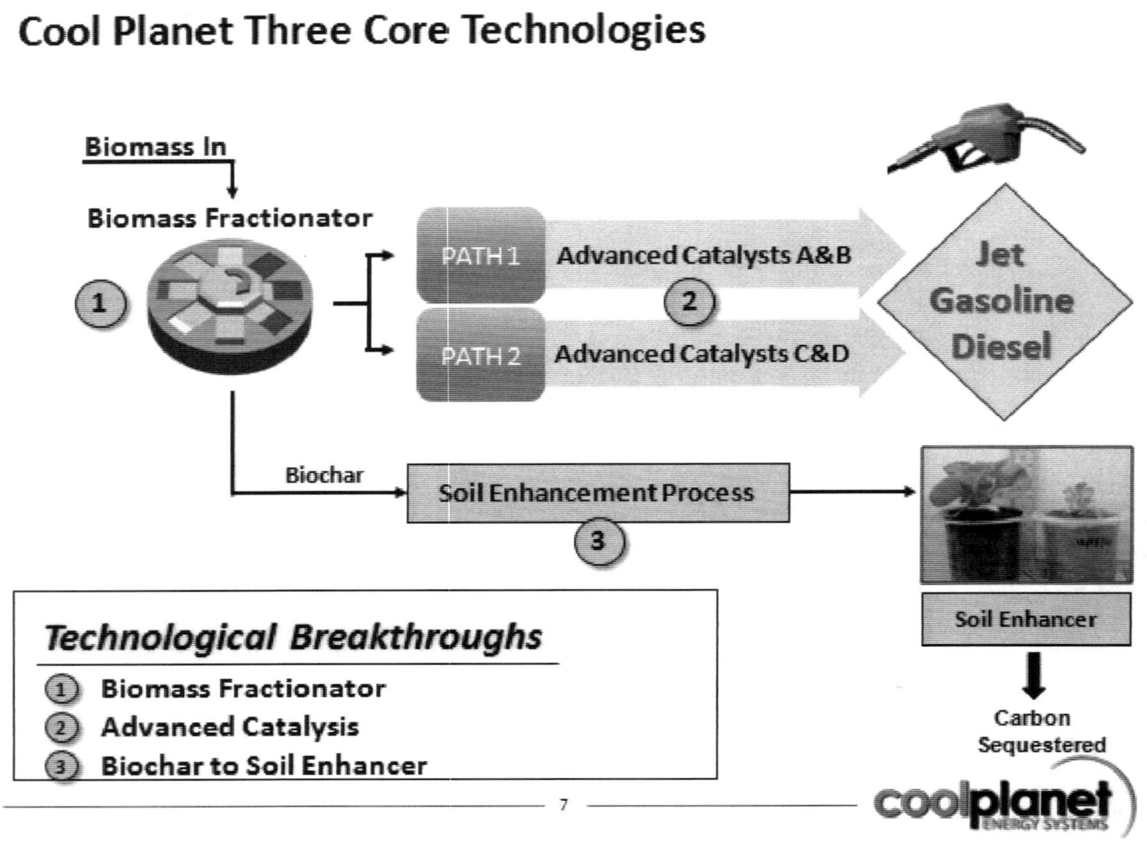

Technology of the Biomass fractionator

Ground biomass is transported by a commercial, mechanical transport system to a load hopper that pre-stages the biomass for the fractionator. Methods of introducing biomass into a processing station can include introduction via conveyor belts, hoppers, and/or pulverizers. The biomass is fed to the top of a sliding gate valve above a load station, and a precise amount of biomass is placed in a moving tray that holds the biomass for the next stage of processing. The biomass moves to the 1st heating and processing stage where the function is to drive off entrapped water, and to perform 1st step of de-hydration of the biomass while mechanically compacting it for more effective heat transfer at later stages of processing. The processing stages rotate or use a shuttle drive from step to step where the biomass is progressively heated and mechanically compacted from an initial stage temperature of 150 degrees C. As each of the following stages heat and compress the biomass, it will outgas commercially viable bio-intermediary compounds that are then taken off to catalytic reaction chambers. At each successive processing stage, the biomass is heated to a higher temperature profile and the material is compacted to a thinner layer with increasing level of elemental carbon in the remaining biomass. The processing stations thus produce volatile and non-volatile components when heated. Because the material can be heated within a relatively narrow temperature range, the non-volatile component (in the form of partially formed char) has additional volatile components embedded therein, which are extractable and processed in subsequent processing stations. The output of these processing stations is connected to an array of catalysts as described in the next section.
The final step or stage does not compact but controls the final biochar product output quantity and quality.

Advanced Catalytic Reaction Arrays

The Cool Planet system has a programmable number of processing stations, connected to an array of catalysts; a means for subjecting biomass within these stations to programmable temperatures and increments to produce a volatile and non-volatile component; a means for incrementing the non-volatile component temperature to further produce additional volatile and non-volatile components; and a means for subjecting the volatile components generated in each station through a series of catalysts. The catalysts involved are selected from the group of dehydration catalysts, aromatization catalysts, and gas-upgrading catalysts to produce different types of renewable fuel. The chemical compounds from biomass decomposition are routed to these distinct catalyst chains to produce products containing renewable, pure HC fuels and other value added products. Biomass is typically comprised of a wide array of compounds classified within the categories of cellulose, hemicelluloses, lignin, starches, and lipids. These compounds go through multiple steps of decomposition when subject to the biofractionation "pyrolysis" process. For example, hemicelluloses comprise C5 sugars such as fructose and xylose, which yield furfural and hydroxymethylfurfurals upon thermolysis. The latter compounds can be further converted to fuel intermediates furan and tetrahydrofuran. The narrow temperature windows experienced within a processing station allows for the collection of these useful intermediates.

The processing stations are connected to a series of catalyst components comprising a dehydration catalyst, an aromatization catalyst, and a gas-upgrading catalyst. The main catalysts are proprietary modified ZSM-5 catalysts. The output from each catalytic array when cooled is comprised of volatile gases, renewable fuel and water. The volatile gases are then programmed to pass through subsequent catalytic columns. Depending on the temperature (300C to 600C) of the processing stations and the biomass composition, the volatile components from any given processing station can be channeled through one or more catalysts using program selector switches.

Biochar

The last step as heat energy transfer is completed to the biomass is the production of biochar (BMF char). The BMF char remains after the previous processing steps, and the physical characteristics of the char will differ depending on the starting biomass material, which can include any of the above-identified cellulosic materials such as wood, grasses, energy crops, and agricultural waste. Different biomass feedstocks are expected to produce different types of BMF chars, varying in porosity and other physical characteristics. The biomass feedstocks can be fed individually or as mixtures of different feedstocks to produce chars containing different physical characteristics. After the BMF char is formed, a last processing step is to transfer the BMF char out of the reaction chamber for a subsequent storage or filling into a bio-char reactor for subsequent optional activation. The transfer may be accomplished via any number of mechanical means, including a press bar outfitted with a scraping knife. The Company has developed a novel process for rendering the BMF char suitable as a soil growth agent. This process comprises several steps including creating the BMF char, expelling detrimental agents within the BMF char, rendering the internal surface area of the BMF char hydrophilic, and adding suitable nutrients and microorganisms to the BMF char to promote plant growth. In addition to acting as a soil enhancing agent, the BMF char can sequester carbon for long periods of time.

Biochar Co-product – Return Carbon to the Soil

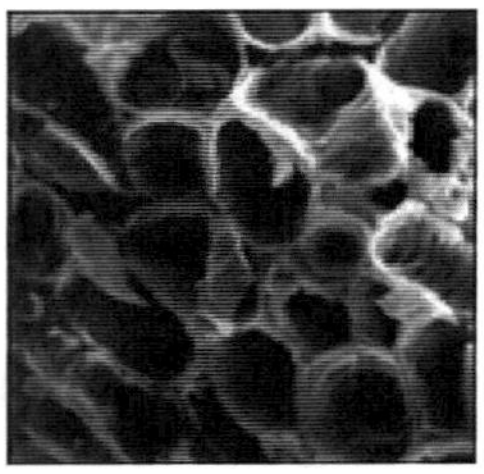

Biochar is a highly porous form of carbon produced by the selective heating of biomass, in the absence of oxygen.

Biochar is very promising as a soil amendment that can retain water and inputs in the soil. Biochar can help maintain a nutrient rich environment and increase productivity of a wide variety of crops.

Drop in Hydrocarbon Fuel Results

Upon completion of catalytic conversion, the devolatilized organic components convert to various liquid fuels such as biodiesel, hydrocarbons, BTX aromatics, jet fuel, light hydrocarbons, gasoline, diesel, methanol, and DME. The organic chemicals can also be useful on their own as co-produced chemicals. By contrast, typical pyrolysis processes do not exhibit a clear volatilization profile such as this. 3rd party testing by independent laboratories shows the compounds consistent with a high octane gasoline as shown in the next graph.

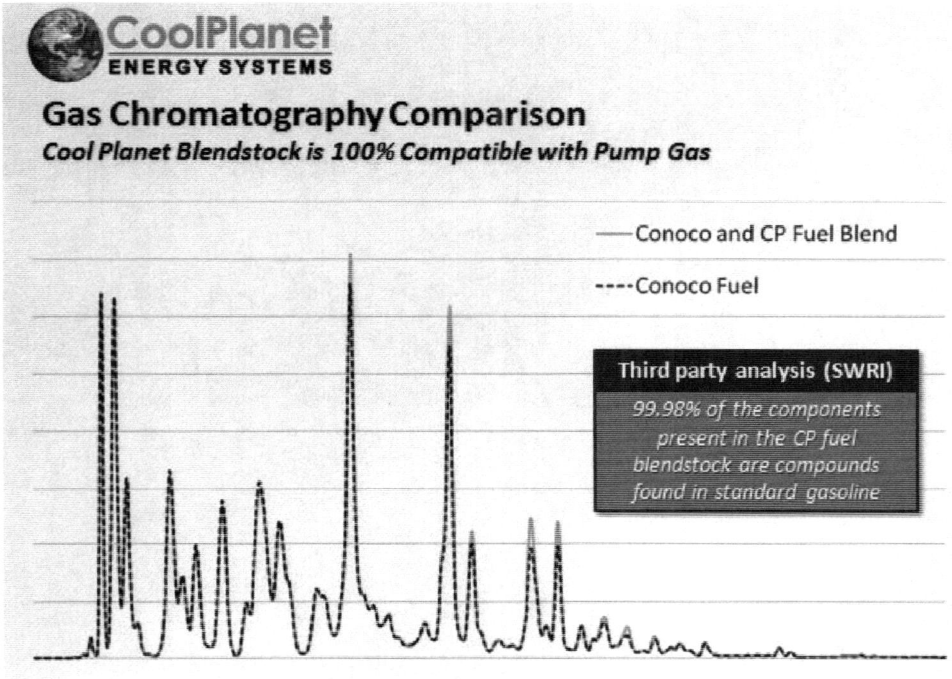

Mass Production of Distributed Units

The Cool Planet refinery equipment is modular, with fractionator and catalyst equipment connected as pairs of processing equipment. This is transported to various biomass locations where the modules are placed within a 30 mile radius of the biomass to be processed. Capex is lower by a factor of 10 compared to existing biomass refineries.

The modular refining units will be manufactured and assembled at a central plant manufacturing facility. Coordinated team will then roll out the modular facilities across the US.

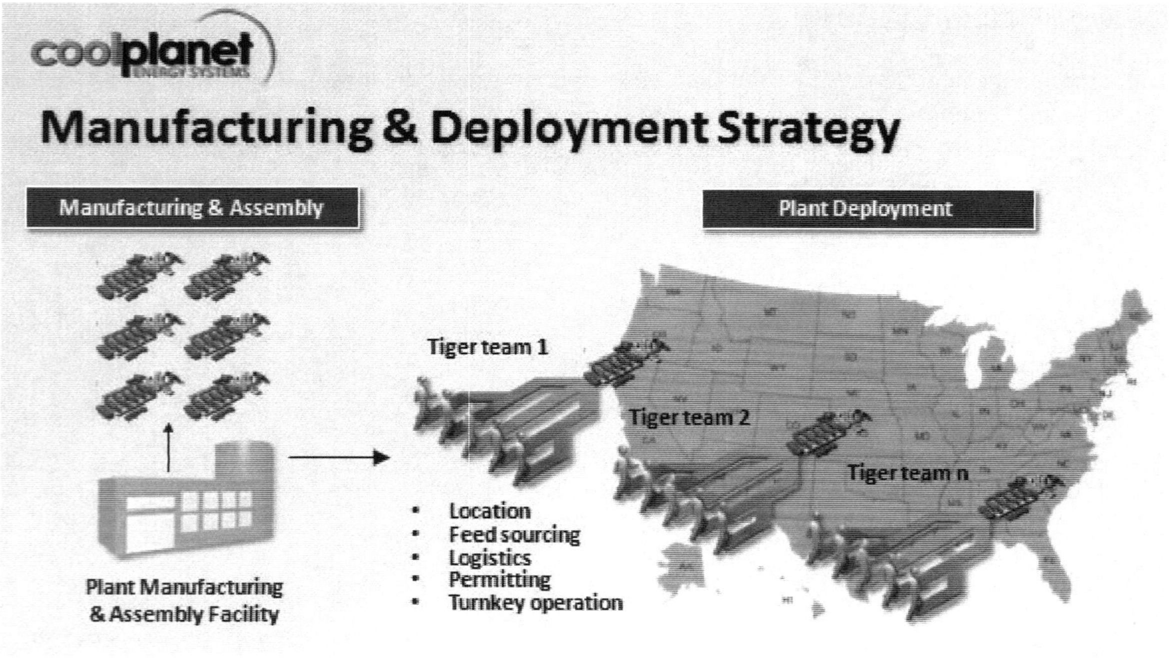

US and Global Roll-out Plan

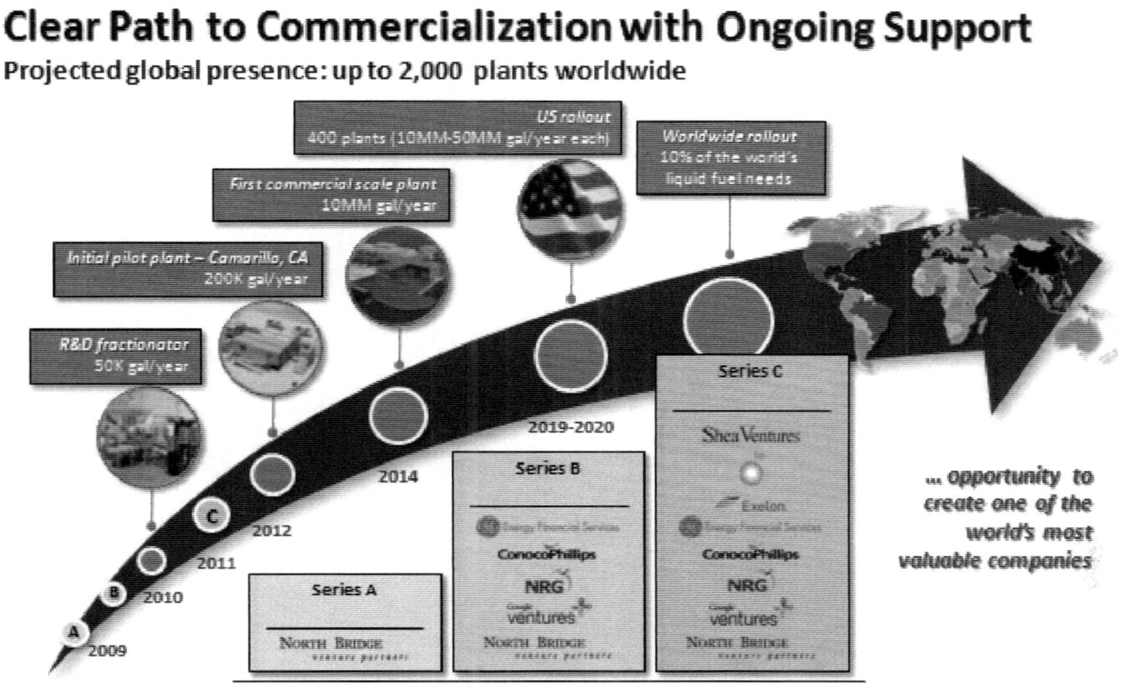

AUTHOR INDEX

AUTHOR INDEX